GW01465439

INVEST IN LIVING

HOME PLUMBING

by

Tom Pettit

EP Publishing Limited

Acknowledgements

Cover Photograph:
  Chloride Shires Ltd
In addition the author wishes to thank the following manufacturers and associations for the information which they have so willingly provided:
  Bartol Plastics/Plastidrain Ltd/
    members of the Plastics Division of
    Hepworth Ceramic Holdings
  British Standards Institution
  Camping Gaz International
  Conex Sanbra Ltd
  Deltaflow Ltd
  National Water Council
  Rawlplug Co. Ltd
  Rabone Chesterman Ltd
  Spear & Jackson Ltd
  Stanley Tools
  Wolf Electric Tools Ltd
  Yorkshire Imperial Metals Ltd
  Yorkshire Water Authority

The *Invest in Living* Series

All About Herbs
Fruit Growing
Gardening under Protection
Getting the Best from Fish
Getting the Best from Meat
Growing Unusual Vegetables
Home-baked Breads and Scones
Home Decorating
Home Electrical Repairs
Home Furnishing on a Budget
Home Goat Keeping
Home Honey Production
Home-made Butter, Cheese and Yoghurt
Home-made Pickles and Chutneys
Home Maintenance and Outdoor Repairs
Home Poultry Keeping
Home Rabbit Keeping
Home Vegetable Production
Home Woodworking
Improving Your Kitchen
Meat Preserving at Home
101 Wild Plants for the Kitchen
Wild Fruits and Nuts

About the Author

Tom Pettit has taught woodwork, metalwork and technical drawing for thirty years and is at present Head of the Craft and Design Faculty at Aireville Secondary School, Skipton. He participated in the School's Council Craft and Design Curriculum Development Project from which has developed the School's national reputation for Community Service in craftwork. He is a founder member of the Craft Teachers' Centre in Burley in Wharfedale which was the first of its kind to be set up in the country.
Tom Pettit has also written *Home Maintenance and Outdoor Repairs* in the Invest in Living series.

Copyright © Tom Pettit 1978

ISBN 0 7158 0464 2

Published 1978 by EP Publishing Ltd, East Ardsley, Wakefield, West Yorkshire WF3 2JN

This book is copyright under the Berne Convention. All rights are reserved. Apart from any fair dealing for the purpose of private study, research, criticism or review, as permitted under the Copyright Act, 1956, no part of this publication may be reproduced, stored in a retrieval system, or transmitted in any form or by any means, electronic, electrical, chemical, mechanical, optical, photocopying, recording or otherwise, without the prior permission of the copyright owner. Enquiries should be addressed to the Publishers.

Printed and bound in England by The Scolar Press Ilkley West Yorkshire

Contents

Introduction

The installation of a complete bathroom suite, including all the necessary plumbing and the fitting of a fireback boiler in a house previously without a bathroom is within the author's experience. Many enterprising householders must have undertaken this project successfully just as many others have installed their own hot-water central heating systems. Although such achievements are not uncommon, they do necessitate considerable preplanning. Research is required to ensure the efficiency of the completed work; care taken to avoid breaching local authority requirements; knowledge of the structure of the building because it is inevitable that walls will have to be drilled and floors taken up; the craft skill and confidence necessary to carry this out, coupled with a considerable financial outlay.

However, these are rather ambitious projects and somewhat outside the brief of this book which is aimed at the householder who wishes to be self-sufficient regarding what can be considered as minor domestic plumbing, thereby avoiding the need to seek professional help for all but major tasks.

In the belief that basic knowledge is of fundamental importance, the opening chapters discuss the domestic water supply in general indicating how water is provided by the local water authority, the usual arrangement of the average domestic plumbing system and the few common tools necessary to cope with simple repairs and modifications to it. Up to the 1920s most domestic plumbing was carried out entirely in lead piping which was heavy, unsightly, vulnerable as lead is such a soft metal, and in some circumstances a health hazard. Copper, and more recently plastic pipes have almost entirely replaced lead, but have necessitated the development of new techniques with which to work them. These are included within the text, as are the techniques required for the maintenance of taps and waste pipes. The book also discusses the possibilities of replacing damaged or outdated components, and the installation of the modern shower unit.

Although drainage is normally considered to be the province of the builder, the author is of the opinion that as it is such an integral part of the water system of a house it is proper to include it within the book, and some basic details of care and maintenance are given.

One of the many purposes of the *Invest in Living* series is to encourage the householder to utilise leisure time gainfully. By using this book wisely it is hoped that it will help the reader to do just that and at the same time prevent what are really only minor faults becoming major operations.

As the conditions, and the degree of skill by which any suggested work is carried out, are beyond the control of the publishers or the author, results cannot be guaranteed. However, every effort has been made to ensure that information provided is correct and that within the space available procedures are sufficiently detailed.

If any modifications to the existing plumbing installation are being considered, either to the pipework, or by the addition of other fitments there could well be regulations which you must observe. To avoid infringement of your regional water authority requirements you are advised to make the necessary enquiries.

1 The Supply of Domestic Water by Local Authorities

The oceans of the world contain nearly 90 per cent of all the water on earth, the remaining 10 per cent being in rivers, underground in the pores and fissures of rocks and as moisture in the atmosphere. All this makes up what is a fixed volume of water which circulates by natural climatic processes in the so-called hydrological cycle. Water evaporates from the oceans mainly due to the heat of the sun, the resulting water vapour forming clouds which are blown over the land. When conditions are suitable rainfall is precipitated from them, finally making its way back to the sea again in streams and rivers. This is surface water, or it may percolate through the underlying rock strata seeping back to the sea as underground water. Water vapour is also transpired into the atmosphere by plants and trees. Fortunately, since water is essential to our lives, the water authorities have the necessary technical knowledge to utilise this continuous natural cycle to meet our ever-increasing demands. It is estimated that a total of 22 million litres (approximately 5 million gallons) per day is required to meet the domestic, industrial and agricultural needs of a population area of 100,000 people. Approximately 20 per cent of the population lives in the area north and west of a line joining the Severn Estuary to the Wash, the remaining 80 per cent being in the south and east. Geologically the ground to the north and west is mountainous and consists of older and harder rocks than the flatter south and east. Rainfall is therefore higher in the north and west, approximately 1,800 mm (70 in) per year as opposed to 500mm (18 in) in the south east, because our prevailing winds are moisture ladened south-westerlies from the Atlantic Ocean which precipitate heavy rain over the high ground. *(Fig. 1.)*

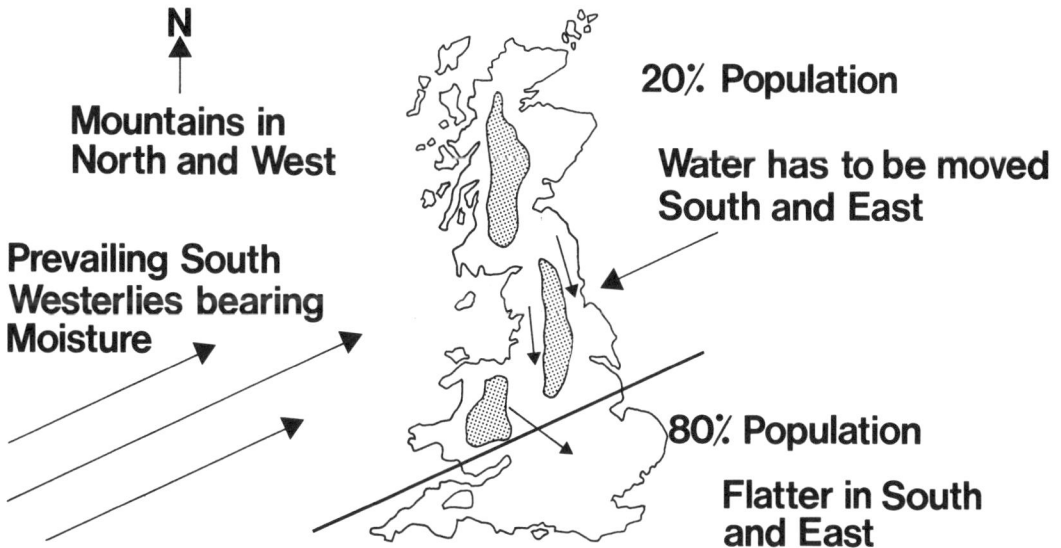

N

Mountains in North and West

Prevailing South Westerlies bearing Moisture

20% Population

Water has to be moved South and East

80% Population

Flatter in South and East

Fig. 1.

At present all water supplies depend upon rainfall, although fresh water can be distilled from sea water by a very expensive process. The deep valleys between the mountains of the west and north are suitable for storing the run-off rainfall by constructing dams across them to create reservoirs. In the flatter south east rivers are utilised and underground resources are tapped with boreholes. *(Fig. 2.)*

There is usually a vast natural store of underground water so that a regular supply can be pumped from wells and boreholes, whereas the run-off from streams and rivers can be variable—heavy in winter and light in summer. Because of this there is the need to build impounding reservoirs in which to store the water. An alternative surface supply may be from a river, the water being pumped to a nearby reservoir. Reservoirs at or near the source of water are generally for long-term storage, while service reservoirs are near to where the water is to be used and are short term.

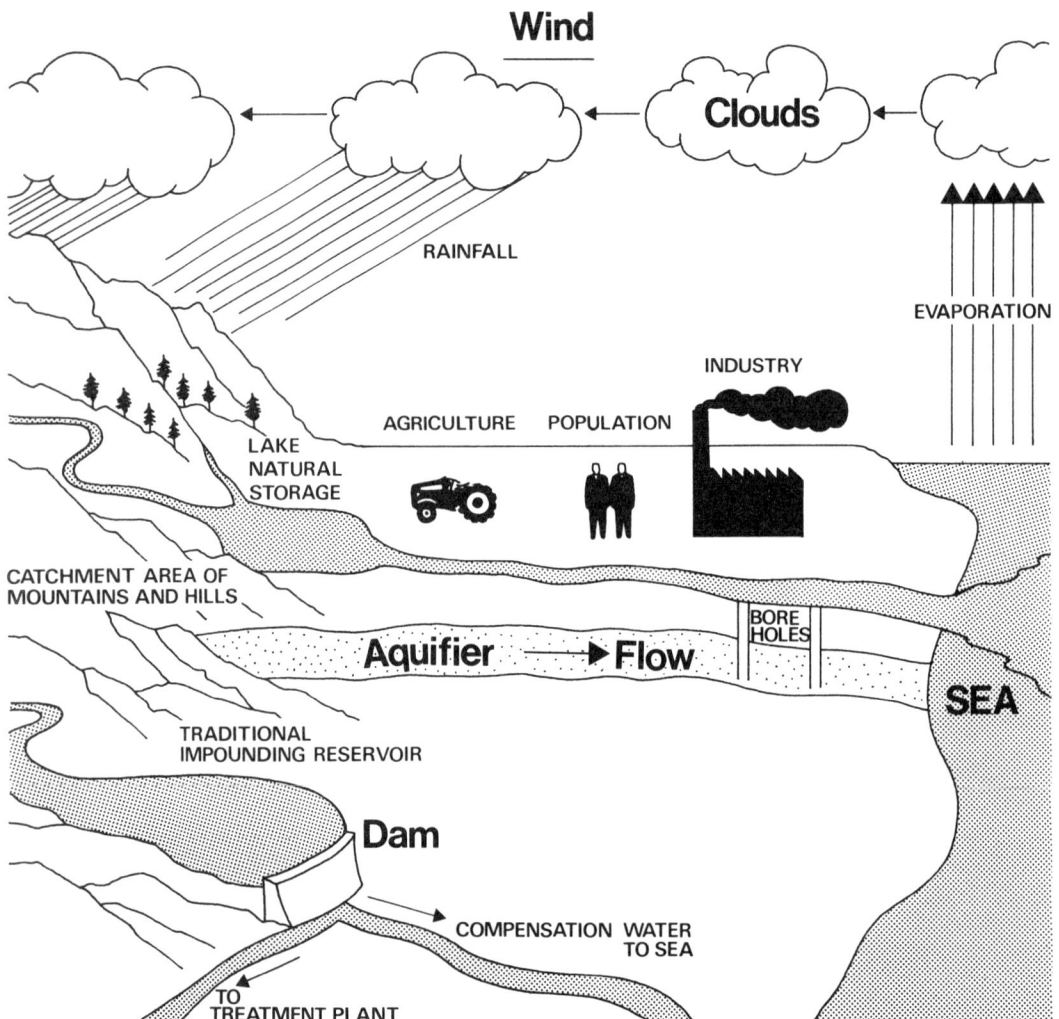

Fig. 2. The hydrological cycle

By Act of Parliament the consumer is entitled to a wholesome potable water which can be defined as being: clear, colourless, tasteless, and odourless, free from sediment, free from living creatures, free from bacteria, free from toxic substances and not corrosive to metals, especially lead. This means that the water must be suitably treated before it is supplied to the customer.

Large pipelines or aqueducts convey water from the works at its source. The water may be gravity fed from a high level to a lower one, or it may be pumped if there is not sufficient 'fall'. Depending upon the geological formation of the area which the water is to serve it is fed into service reservoirs at ground level, or into raised water towers, from where the water mains distribute it to the customer. (Fig. 3.)

The water is made potable at the water-treatment works by screening, straining, sedimentation, filtration and sterilisation. Floating matter and some suspended particles are removed by screens, very fine residues sometimes being removed by microstrainers. Both horizontal flow and vertical flow sedimentation tanks are used to settle out further suspended matter followed by filtration. This is sometimes assisted by chemically coagulating the suspended matter. Any remaining impurities and most of the bacteria are removed in sand filters operated by gravity or under pressure. The water is finally sterilised with chlorine to ensure that all bacteria have been killed.

So called 'hard waters', which on account of their mineral composition do not lather easily, may be 'softened' at the works by the addition of lime or soda. Alternatively, softening may be by 'ion exchange' either at the works or by a domestic water softener. It is interesting to note that synthetic detergents are equally effective in hard or soft water, as these are not dependent on a lather for cleaning.

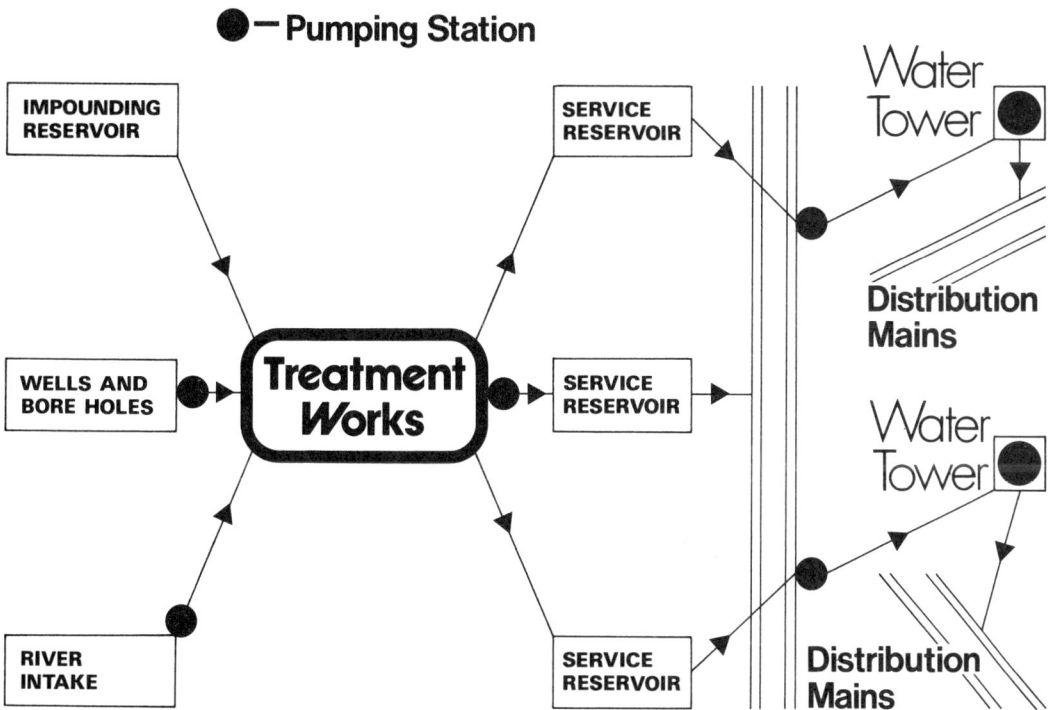

Fig. 3. Water sources, treatment and distribution

2 The Domestic Plumbing System

The average household uses about 450 litres (100 gallons) of water per day for personal drinking, washing and bathing, washing clothes, domestic utensils and the car. It is also used for washing floors and paintwork, the preparation of food, flushing toilets and at times watering the garden. Before any maintenance to the plumbing system is considered it is important to know how it works.

Your plumbing system begins at the point where the water leaves the water authority's stopcock. This is usually near to the boundary of the property, and near to the authority's water main in the road. Its purpose is to shut off the supply when repairs are to be carried out, and as a precaution against frost is usually about 760 mm below ground. A turn key (see Chapter 3) is generally required to turn it on account of its depth. Access to the stopcock is by a hinged lid on an iron surface box, the lid having a small recess on the edge opposite the hinges where a lever can be inserted with which to lift it. The supply to a number of houses may be controlled by one stopcock if the property is old and this should be checked or neighbours may be inconvenienced! Where the service pipe enters the house there should be a second stopcock (often under the sink), or in a block of flats one to each, so that the system can be isolated. If the water pressure in a particular area is low the service pipe may be 22 mm, but it is more usual to be 15 mm for domestic buildings where there is average water pressure. The stopcock works in just the same way as a tap (see Chapter 7) to control the flow, but is designed to be plumbed into a pipeline having either

capillary or compression connections at each end. An arrow on the body shows the direction of flow and this must be observed. If fitted wrongly the water pressure will simply hold the washer down, cutting off the supply. Water authority regulations require that the jumper must be of the loose type so that if the mains supply fails the jumper and washer fall into place (see Chapter 7). This acts as a non-return valve preventing a backflow of water into the mains with the additional possibility of contamination.

The advantage of the second stopcock within the property is in being readily available in an emergency. Imagine the sort of problem you might have on a dark night with the only stopcock outside under several metres of snow and ice!

Once the supply passes the second stopcock it is known as the 'rising main' and then becomes part of a 'direct' or 'indirect' system. Whichever system it feeds, however, water authorities require that a supply point must be taken from it where water is required for cooking and drinking; this is generally in the kitchen so it is usual for the mains supply to enter the building just under the kitchen sink.

Direct System

Direct systems are more usual in older properties, all cold draw-off taps being fed directly from the rising main. *(Fig. 1.)* There are advantages and disadvantages to this system.

Advantages
■ Potable water can be had from any cold tap.

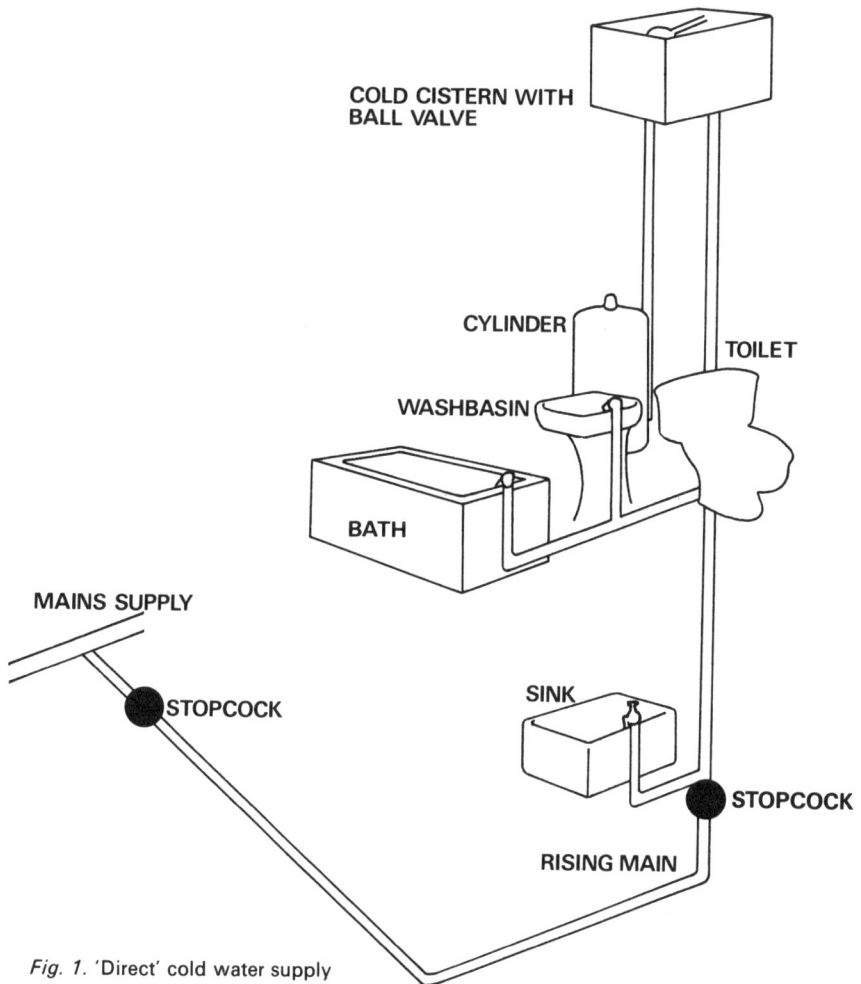

COLD CISTERN WITH BALL VALVE

CYLINDER

TOILET

WASHBASIN

BATH

MAINS SUPPLY

STOPCOCK

SINK

STOPCOCK

RISING MAIN

Fig. 1. 'Direct' cold water supply

- The cold cistern to supply the hot-water cylinder is at a lower level, usually in the airing cupboard, and so is not subjected to freezing.
- Being in the airing cupboard dirt is less likely to fall into it.

Disadvantages

- All taps and ball valves are at mains pressure which is quite high and are therefore subjected to more wear.
- High pressure ball valves on cisterns and toilets are more noisy, although many have silencer tubes fitted (see Chapter 7).

- There is more chance of noise in the pipes due to 'water hammer'. This is often due to the ball valve bouncing before it finally cuts off, because of the greater pressure. The noise can be transmitted throughout the system.

- Because it is possible to draw off cold water directly in this way from several points at once it leads to greater demands being made upon the mains supply.

- Increased danger from 'back siphonage' if mains pressure falls.

Indirect System

It is more usual in modern properties to employ the indirect system (*Fig. 2.)* where there are no direct cold draw-off points other than the one in the kitchen. The rising main then feeds a large capacity storage cistern in the underdrawing or roof space, a suitable size being 320 to 360 litres (70 to 79 gallons) approximately. This cistern then supplies cold water to all other parts of the system. It is placed as high as can be to give as big a head of pressure as possible at each draw-off point as they are now only gravity fed.

Because of this, indirect supplies to the bath at least require to be run in a 22 mm pipe to obtain an adequate flow, although 15 mm to washbasins is usually satisfactory.

Advantages
- A quieter system as the supply cistern is in the roof space and other cisterns and cold taps are fed from it at a lower pressure.
- The large cistern provides a reserve store of water for other than drinking purposes should the mains supply fail.
- The lower pressures under which the system operates produce less wear and tear on the pipes and fittings.

Disadvantages
- There is only one point in the building where really potable water is available – the kitchen.
- The supply cistern, sited as it is in the roof space, may collect dirt if it is open at the top.
- It is also exposed to possible frost damage particularly in view of ceilings being insulated. The cistern must therefore be adequately lagged (see Chapter 14.).

With either system it is usual for 'instantaneous water heaters' – geysers – to be supplied from the rising main since the hot water they supply may be used

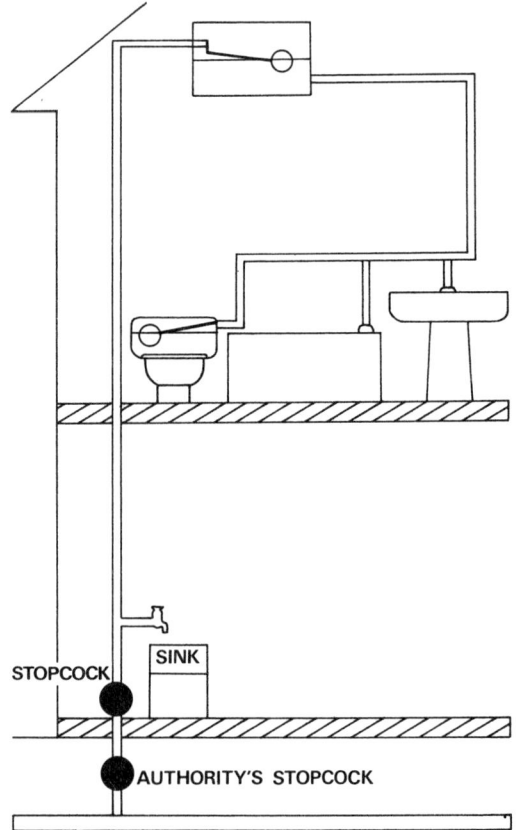

Fig. 2. 'Indirect' cold water supply

in the preparation of food and drinks, and for washing up, among other things.

Larger amounts of hot water may be produced from a number of different energy sources – the electrically powered immersion heater, the traditional fireback boiler, or a variety of solid fuel, gas or oil-fired independent boilers. Very often these also provide hot water for central heating, a pump being placed in the circuit to ensure rapid and even circulation. Domestic hot water from each system is gravity fed and therefore at a comparatively low pressure so 22 mm pipe is used to ensure an adequate flow to baths where large quantities of hot water are required quickly.

Again, there are two systems, the 'direct'

and 'indirect'. In each case the hot water is collected in a copper storage cylinder below the cold-water cistern in an airing cupboard. Cold water is fed from the cistern to the bottom of the cylinder through a 22 mm pipe which in the direct system feeds the boiler. Hot water from the boiler then rises and collects in the top of the storage cylinder. This can be drawn off at the hot taps, sufficient 'head of water' being provided by the cold-water cistern through the cold feed pipe and the pressure of the expanded hot water. As this hot water is drawn off it is replaced by the cold water flowing in at the bottom, and by circulation between the cylinder and boiler, further hot water collects at the 'crown' of the cylinder. *(Fig. 3.)* This system provides domestic hot water only.

In districts where the water is hard (see Chapter 1) and may lead to scale or 'fur' being deposited in the boiler and pipes, an 'indirect' cylinder should be used. This separates the 'primary' water, i.e. the water circulating through the boiler, from the 'secondary' which is drawn off for domestic use. The primary water heats the secondary by transmitting heat through the walls of the inner cylinder, copper being a good conductor of heat. In this way the primary water circulates continuously and as it is not drawn off there is only a small initial deposit of scale with little further increase. A small cold-water feed cistern is necessary to maintain the supply of water in the primary, making

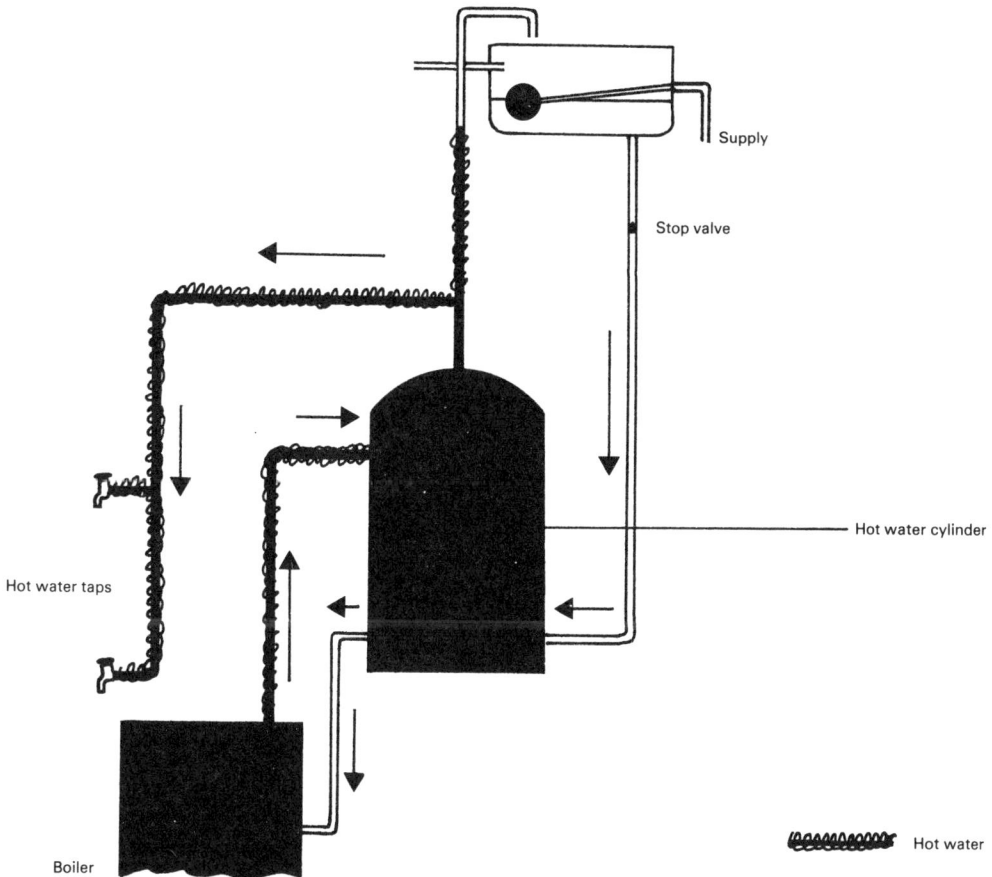

Supply

Stop valve

Hot water cylinder

Hot water taps

Hot water

Boiler

Fig. 3. Hot water – direct system

good any loss because of evaporation, or from steam caused by boiling. Direct and indirect systems require expansion pipes from the top of the cylinder reaching over the edge of the cold-water cistern(s). Any surge of expanding hot water is then contained. The end of the expansion pipe must not touch the surface of the cold water or water may circulate between the cistern and cylinder.

Radiators for central heating can also be on this scale-free circuit, the water not being able to contaminate the domestic supply. A cylinder with a calorifier, i.e. a coiled copper tube, which replaces the indirect cylinder may be used as an alternative. *(Fig. 4.)*

Heat from the hot water passes to the metal radiator, and by radiation and convection heats the air which circulates in the room. Recommended temperatures are 16°C (60°F) for bedrooms and 21°C (70°F) for other rooms. Adequate radiators must also be positioned in halls and staircases to prevent a heat loss from the rooms.

Radiators may be of cast iron or pressed steel, the amount of heat they radiate depending on their surface area and the temperature of the water. At one time, particularly in public buildings, radiators were always matt black, it being generally acknowledged as the best radiating surface, this is now ignored in favour of modern colour schemes. Radiators are so designed to have the maximum surface area, either in the form of elliptical pipes or ribbed panels.

Heated towel rails are usually plumbed into the domestic hot-water system rather than into the central heating system since they are required to be in continuous use.

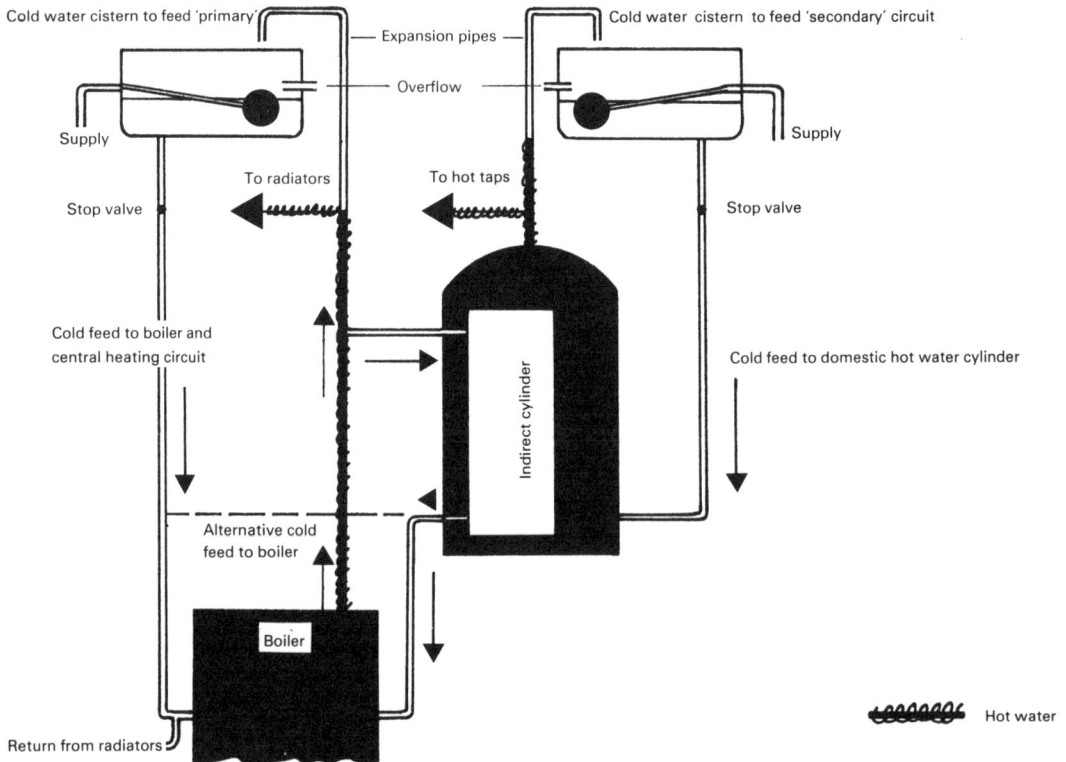

Fig. 4. Hot water – indirect system

3 Plumbing Tools and Sundries

Several of the tools used for plumbing are of the general purpose variety, being necessary items of equipment for other aspects of home maintenance. Naturally there are many others designed specifically for plumbing to ensure the efficient manipulation of pipework and the installation of fittings. The skilled plumber, as any other skilled craftsman, knows the tools of his trade and is thoroughly conversant with their use, knowing that without them the work cannot be carried out effectively. This more than holds good for the enthusiastic amateur, as it would be folly to embark upon even the simplest of plumbing projects without adequate tools. He or she should study the proposed work carefully, making sure that the requisite tools are available and if necessary practise with them on surplus materials beforehand. Most modern domestic plumbing installations are run in copper pipe, or a combination of copper and plastic. Plastic is suitable for waste-water pipes, and may also be used for the cold-water supply (see Chapter 4). It is sound advice to trace the pipework in your home, so that being familiar with it you have the tools available to deal with any emergency which may occur. Probably the most elementary of plumbing repairs is the renewing of a tap washer for which only the minimum of basic tools are required, but with experience more ambitious repairs and modifications may be undertaken for which specialist tools will be necessary.

With this probability in mind the following list is suggested. They will allow simple work to be carried out with ease, and also be the foundation of a tool kit by which the majority of domestic plumbing work can be accomplished.

1. Screwdrivers – general purpose for slotted screws x 100, 150 and 250 mm.
2. Posidriv screwdrivers – many modern fittings employ screws with heads of this type. No. 2 fits screw gauges 5 to 10 and is suitable for most household tasks. Both types being generally shockproof will withstand considerable rough use and have plastic handles which are impervious to water.
3. Pliers – combination type x 175 mm – those with insulated handles are much more comfortable to grip, but care must be taken to avoid melting the plastic cover if working with a blow torch. Will cut wire and grip small tubes, thin sheet metal and small nuts.
4. Pliers – gas – the jaws are designed to grip and turn pipes, and other cylindrical work, but the teeth are coarse and will cause marking if undue force is used. One handle has a screwdriver end which can be used as a screwdriver or a lever; the end of the other handle is conical and can be used for flaring the ends of small-bore pipes.
5. Pliers – universal gland nut – slim and adjustable to several positions. The jaws are serrated, the teeth so formed for gripping cylindrical shapes firmly. As with all pliers there is the danger of marking the work, so if this is a matter of some concern, e.g., chromium plated or stainless steel fittings, they must be protected by a cloth or heavy self-adhesive tape.

There are many adjustable wrenches, and the following are some of the different types used in plumbing:

6. Wrench — mole (from the original advert, 'your third hand') — the advantage is that they may be clamped onto the work and will stay in position until released, leaving both hands free to do other work. The capacity of the jaws is adjusted by the knurled screw at the end of the handle and the tool can then be locked in position by pulling the handles together.

7. Wrench — adjustable open-ended (often referred to as an adjustable spanner) — these are used on hexagonal or square nuts, and the mechanisms of taps and valves. Their size is measured by length, for general purposes 310 mm is adequate, 205 mm for smaller work. The larger size has a bigger capacity. Note how the wrench is fitted to the nut so that the applied force is towards the moving jaw. The nut should be as far into the jaws as possible which in turn should be adjusted to fit firmly round it. Used in this way the jaw will not open accidentally, resulting in the wrench slipping off with possible damage to the work and hands.

8. Wrench — monkey — used for the same kind of work as the adjustable open-ended wrench, but as it is somewhat thicker is not quite so convenient if space is limited. Care should be taken how the wrench is fitted to the work and in which direction the force is applied — see *Fig. 8*. Similar sizes as for the open-ended wrench are suitable.

9. Wrench — 'Footprint' — adjustable pipe — basically for pipework only, the turning force applied locks the wrench firmly to the work.

10. Wrench — 'Stillson' — pipe — used for heavier work on iron pipes on which the hardened steel jaws bite and grip. The movable jaw is pivoted so that as leverage is applied to the back of the handle the wrench grips the pipe. Often two wrenches are required on a job, one to hold the pipe and one to turn the fitting. As they work only in one direction, they must be placed in opposing positions. They are most efficient if the jaws are so adjusted that they bite on the work at their centre point. These are tools which are much too heavy for work on brass nuts or light tubing, particularly if it is polished or with a chromium-plated finish. The teeth would certainly leave marks and there would be the danger of collapsing the tube.

11. Wrench — strap — for coping with the light work (referred to above) — light tubes in copper, brass or plastic, or soft lead pipes can be handled without damage to them. The strap is looped round the pipe in the opposite way to the direction of rotation, the end slipped through the slot below the handle and drawn tight. Pulling the handle further tightens the strap so the pipe is gripped and turned. The strap wrench is also ideal for removing the 'easyclean' cover from taps if they prove to be very tight (see Chapter 7.) The chain wrench operates in a similar way, being for really heavy use on large-diameter iron drain or steam pipes.

12. Wrench — basin — specially designed to fit the backnut below washbasin taps where because of curvature of the basin and its proximity to the wall there is very limited space.

Other essential tools are as follows:

13. Plumber's vice — special vices are available, designed specifically to grip pipes. This has a very limited application however, and the new multipurpose 'Lockjaw' vice is so designed that it will meet all the forseeable needs related to domestic plumbing, and many more besides. Irregular work can be held firmly

without damage in self-adjusting jaws. These are interchangeable with optional rubber-faced jaws which will grip polished work without scratching or deforming it.

14. Steel tape – flexible steel tape of at least 2 m length is necessary even if only the most modest work is proposed. It is invaluable for checking the position of fittings and measuring pipe lengths. The tape should be fitted with a 'true zero hook' so that accurate 'hook over' or 'end on' measurements are possible.

15. Hacksaw – junior – ideal for light work and cutting small-bore pipes, as it can be used in one hand leaving the other free.

16. Hacksaw – 250 mm length – for heavier work. Fine-toothed blades should be used for sawing pipework to ensure smooth cutting. By the use of a simple jig, pipes may be cut squarely to length (see Chapter 5).

17. Saw – general purpose – the hardened teeth make the saw suitable for cutting wood, metal or plastic. Ideal for rough sawing in wood where nails or plaster work may be present.

18 Pipe cutter – an alternative to the hacksaw. Clamps onto the pipe which it cuts accurately as the pressure is increased. The pipe cutter illustrated can be fitted with a reamer attachment for removing burrs from the bore of the pipe.

19. Bending springs – these are used to maintain the round section of copper tube and are about 600 mm long. Internal and external types are available, the former being most common, and are tapered with a loop at the end to which a wire can be attached for easy withdrawal. With the spring in place, the tube can be bent over the knee.

20. Files – assorted – the 150 and 250 mm sizes as illustrated will be adequate for most work. The woodwork rasp is useful for the quick roughing down of wood at corners round which pipes are to be run.

21. Hammer – general purpose – the engineer's ball-pein hammer will be found satisfactory. The flat pein can be used for heavy work, the ball for lighter use and for tapping up bursts in pipes – particularly lead – before repairs are made. The head is forged and hardened by heat treatment, the shaft made of ash or hickory which is weatherproofed by the manufacturers. 455g is ideal for general use, but for heavier work on brick or stone use 905g.

22. Chisels – Plumber's wood – these are specially designed and as they are made entirely of metal can be used with the hammer, a mallet being unnecessary. Recommended sizes are from 12–32 mm.

23. Chisels – cold – a wide range of widths and lengths are available, 150 x 13, 230 x 19, and 255 x 25 mm are recommended. The one illustrated is of hexagonal section steel and can be file-sharpened.

24. Rawldrill – toolholder – for use with the hammer – a number of 'Rawldrills' are made to fit. They will be required when walls have to be plugged to carry fitments, pipe clips, etc.

25. Rawlplugs – plastic ones are recommended as they are impervious to water and decay; the plug must be set below the level of the plaster so that its full length is in the brickwork.

26. Brace – 150 mm sweep. The small radius sweep is advisable because holes for pipes have often to be drilled in areas where space is limited. It should also be a ratchet brace for the same reason.

27. 'Jennings' type auger bits – sizes 13, 19 and 25 mm. These bits will drill accurate holes through timber in any direction.

1

2

3

4

5

6

7

10

11

12

14

13

15

16

8

9

17

18

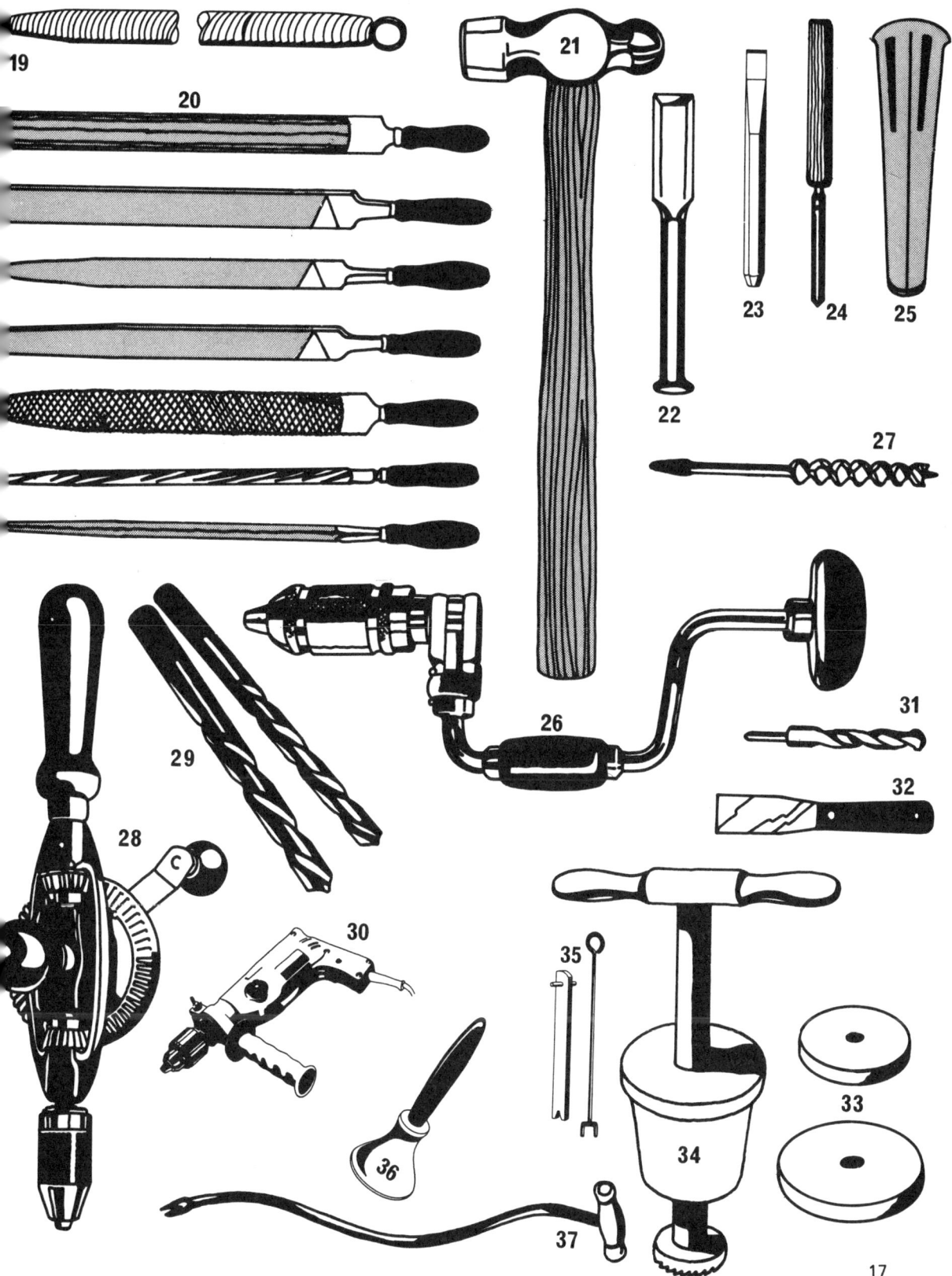

19

20

21

22

23

24

25

26

27

28

29

30

31

32

33

34

35

36

37

17

28. Hand drill — capacity 8 mm for use with metalworking twist drills, and possibly masonry drills. Light drilling in metal and wood, heavier work in brick for rawlplugs.

29. Metalworking twist drills — for general purpose use in wood or metal — diameters 3, 5, 6 and 8 mm.

30. Rotary percussion drill — heavy duty work in concrete, brick and stone. The hammer action makes the drilling of holes in these materials very much easier. The drill illustrated is two-speed with maximum capacity of 13 mm in concrete, 19 mm in masonry, 13 mm in steel or 38 mm with hole saw, and 22 mm in hardwood. It can also be used as a plain drill by disengaging the percussive action.

31. Masonry and percussive drill bits — these have tungsten steel tips for cutting into walls and floors of concrete, etc. Once the cutting action has been established the hole should be completed as quickly as possible to prevent overheating of the bit. It should be withdrawn from time to time to clear waste from the hole.

32. Putty knife — this is useful for applying putty, e.g. to bed a washbasin, and dressing off the surplus. A stout pocket knife with a strong pointed blade is also useful for cutting tape or string and, at times, removing washers.

33. Washers — for draw-off taps and stopcocks; modern hard rubber composition washers are best. Use 19 mm diameter for 15 mm sink and basin taps, and 25 mm diameter for 19 mm bath taps; 10 mm or 13 mm diameter washers are necessary for ball valves depending upon their size and type.

34. Tap reseating tool. The washer seat in the body of the tap should be smooth and free from any defects; this can be checked by visual inspection or by running a finger round its edge. If faulty it can be corrected with a seat-dressing tool of which there are several types all working in much the same way. The tool fits into the body of the tap and the T-handle on the stem is rotated, turning the fluted cutter which trues up the seat.

35. Stop tap turn key — these may be of metal or wood, and are essential for turning off the water at the authority's stop tap should this be necessary. The taps are placed approximately 760 mm below ground as a precaution against frost.

36. Force cup — for clearing lavatory pans when obstructed. The tool is held vertically in the bottom of the pan and the handle worked vigorously up and down.

37. Drain auger or snake — a flexible tool, made of spring steel or coiled wire with a small metal handle. Used for removing particularly difficult objects from waste or drainage pipes.

Miscellaneous Items

Modern blowlamps — operated from small butane gas cartridges or larger cylinders are widely available *(Fig. 38.)*. Neat and powerful they have largely replaced those which burn petrol or paraffin. They are easy to ignite, some being fitted with piezo electric automatic lighting for fine flame and plumber's work; they can be conveniently turned on or off during jobs to save gas. Accessory kits are usually available with these lamps making them easily adaptable to other kinds of work, such as paint stripping and hard soldering. The Camping Gaz Cercotub concentrates the heat, the insulating element protecting the immediate surroundings from the direct heat of the flame.

Glasspaper — medium grade $1\frac{1}{2}$ and fine grade 0 — for cleaning copper and brass prior to making a soldered joint. Used in preference to emery cloth, which may be oily, or steel wool — particles of which may find their way into the joint and by rusting ultimately degrade the solder.

Flux — a non-corrosive paste is recommended for copper and brass work, whereas tallow is required for wiping joints in lead (see Chapter 6).

Soft solder — for use on copper and brass, Tinman's solder 50 per cent tin, 50 per cent lead. For use on lead pipe, Plumber's solder 30 per cent tin, 70 per cent lead.

Graphite impregnated string — for repair of glands; as an alternative use soft white string and tallow.

Plumbers hemp or PTFE (polytetrafluoroethylene) — this is tape for winding round the thread of screwed unions before they are tightened up, thus making them completely watertight.

Jointing compound — such as Boss White — used in conjunction with plumber's hemp and also occasionally on the olives of compression joints (see Chapter 6). The reader is also referred to the tool list provided in *Home Maintenance and Outdoor Repairs* in this series. Many of these tools will be found useful by the home plumber particularly where brickwork has to be made good.

Fig. 38.

4 Pipes for Plumbing

Some metal pipes and fittings may be corroded by the supplied water which is then termed 'aggressive', but other waters may deposit a protective coating preventing corrosion. Local bye-laws will prohibit the use of pipes and fittings made from materials known to be unsuitable for that particular area. With this in mind, piping may be of lead, galvanised steel, stainless steel, copper or plastic. If it is known that the water is 'plumbosolvent', i.e. of a type which can take lead into solution, then lead pipe would be prohibited for health reasons. Similarly, galvanised steel pipe would be prohibited where the water causes severe corrosion. If any brass pipe fittings are to be used enquiries should be made as to their suitability, as some waters cause heavy corrosion of brass called 'dezincification' (brass is an alloy of zinc and copper). Polythene pipe is not suitable for hot-water supply lines because it becomes soft with constant heat, resulting in leaking joints and sagging pipes.

Modern pipework is however almost invariably of copper or plastic. Both are used for interior and exterior plumbing: copper is used almost exclusively for domestic central heating, whereas plastic piping is only suitable for cold-water supply, waste-water and soil pipe systems.

Copper tube should be made from non-arsenical copper to BS 1172 (C106) and manufactured to the standards required by BS 2871. It is available as drawn tube' in Half Hard Temper Table X ($\frac{1}{2}$ H) or Hard

Old imperial bore size	Metric outside diameter	Wall thickness			Lengths in m		Recommended bending spring size	
		$\frac{1}{2}$H Table X	O Table Y	H Table Z			$\frac{1}{2}$H	O
	6 mm	0·6	0·8	—		6		
	8 mm	0·6	0·8	—		6		
$\frac{3}{8}$ in	12 mm	0·6	0·8	—	3	6	10·62	10·18
$\frac{1}{2}$ in	15 mm	0·7	1·0	0·5	3	6	13·40	12·74
$\frac{3}{4}$ in	22 mm	0·9	1·2	0·6	3	6	19·97	19·37
1 in	28 mm	0·9	1·2	0·6	3	6		
$1\frac{1}{4}$ in	35 mm	1·2	1·5	0·7		6		
$1\frac{1}{2}$ in	42 mm	1·2	1·5	0·8		6		
2 in	54 mm	1·2	2·0	0·9		6		
	67 mm	1·2	2·0	1·0		6		
	76·1 mm	1·5	2·0	1·2		6		
	108 mm	1·5	2·5	1·2		6		
	133 mm	1·5	—	—		6		
	159 mm	2·0	—	—		6		

Table Z (H) as indicated. Tube from 6 mm to 54 mm is also made in 20 m coils and is fully annealed – Table Y (O). All are suitable for hot and cold-water services, gas and sanitation installations, grade O being particularly recommended for underground supplies and radiant heat systems.

Soft copper tube for 'minibore' central heating systems is also manufactured.

Outside diameter	Wall thickness	Coil lengths
6 mm	0·6 mm	10, 25, 50 m
8 mm	0·6 mm	10, 25, 50 m
10 mm	0·7 mm	10, 25 m
12 mm	0·8 mm	10, 25 m

You will observe that the soft annealed pipe O has thicker walls than the others. H is light gauge, gaining strength from its hardness. Being hard it is non-manipula-tive and is intended for use in straight runs only. All three grades may be jointed with capillary fittings, care being particularly necessary with soft temper tube where it is essential to re-round the ends with a former after cutting and before jointing. Similarly, all grades may be jointed with compression fittings (see Chapter 5), but it is necessary to use type A (non-manipulative) on H, and type B (manipulative) on O. Grades ½H and O in diameters up to 12 mm may be hand-formed as indicated in *Fig. 1*, taking care not to collapse the tube. Tubes being bent on the knee or former without internal support should be eased round gradually and uniformly to the following recommended minimum centre line radii.

Tube	Radius
15 mm	130–180mm
22 mm	250–300 mm
28·mm	300–350 mm

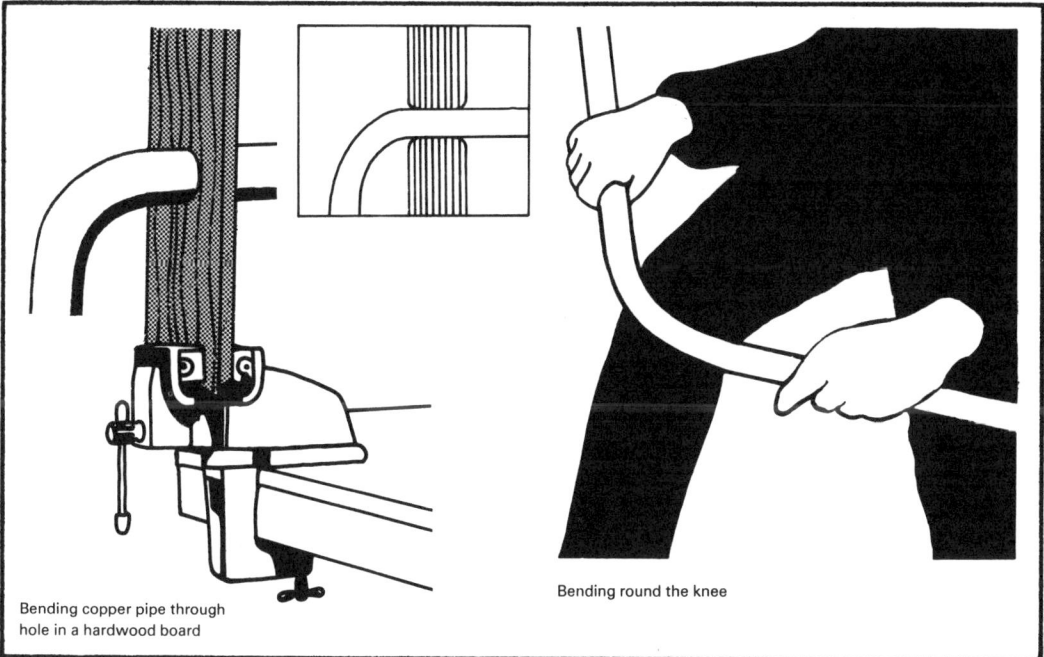

Bending copper pipe through hole in a hardwood board

Bending round the knee

Fig. 1.

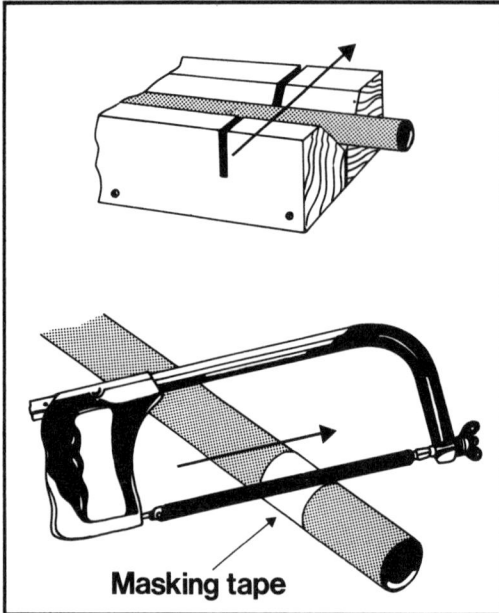

Masking tape

Fig. 2.

The 15, 22 and 28 mm tube should always be supported internally by an appropriate spring when bending. For larger sizes of tube it is necessary to use a bending machine; this can be hired. The tube is held in a groove as it is being bent, so preventing flattening. *Fig. 2.* shows alternative ways by which tubes may be sawn square if a pipe cutter is not available. Any burrs so formed should always be removed with a fine file.

Copper tube, in $\frac{1}{2}$H and O grades, sheathed in seamless polythene and suitable for gas or water service pipes in most corrosive/aggressive situations is available. The colour code for gas is yellow ochre, and green for potable water.

The introduction of plastic piping for plumbing has in many cases made the job for the amateur much simpler, as the materials are lighter and generally more durable. Those in common use are as follows:

Polythene – low density
Very flexible and strong – long pipe runs above or below ground. Does not rot or corrode and is resistant to acids and alkalis. Resists freezing and has good insulating properties. If the water it carries does freeze, because of its elasticity the pipe will expand and then recover as the ice melts. Diameters above 50 mm must not be used below ground as they would be crushed.

The following maximum pressure ratings are based on water at 20°C (68°F):

Low density polythene pipes to BS 1972/67
Class B/Red Mark 6·1 kgf/cm² (0·60 MN/m²: 200 ft.hd: 86·6 lbf/in²).
Class C/Blue Mark 9·14 kgf/cm² (0·90 MN/m²: 300 ft.hd:130 lbf/in²).
Class D/Green Mark 12·2 kgf/cm² (1·20 MN/m²: 400 ft.hd: 173·3 lbf/in²).

High density pipe has greater tensile strength and being more rigid is more suited to withstand higher pressures in large bore pipes, Class C and D.
1 lbf = 4·448 22N

Polyvinyl chloride – PVC
The unplasticised form is generally used for soil and drainpipes. Also domestic cold-water supply systems.

Acrylonitrile butadiene styrene (Abs) and polypropylene
This is used for hot-water waste systems because of its relatively high softening point.

Polythene piping is available in coils up to 150 m in length, pipes for waste-water and overflow systems are usually in lengths of 3 m in the following diameters:

	Mean outside diameter	Nominal inside diameter
Push-fit waste system	34·6 mm	32 mm
	41·0 mm	38 mm
	54·0 mm	50 mm
Solvent-weld waste system	36·3 mm	32 mm
	43·00 mm	38 mm
	56·00 mm	50 mm
Push-fit overflow system	21·5 mm	19 mm

Recommended bores for plastic pipes are: washbasin 32 mm, bath and sink 38 mm and WC 75 mm or 102 mm.

Bends can be made in plastic pipes as follows:

Polythene – low density

Cold bending – mean radius not less than eight times external diameter. Hold with pipe clips at each end of the bend or it will straighten.

Hot bending – insert bending spring, heat in boiling water or fan gently with blowlamp.

Caution – overheating will melt the pipe.

Polythene – high density

As above – mean radius for cold bend ten times external diameter. Hot – with blowlamp, six times external diameter.

In each case allow hot bends to cool before removing spring.

Polyvinyl chloride (PVC) and acrylonitrile butadiene styrene (Abs)

Heat with blowlamp, avoiding the end of the pipe to prevent distortions. With care this can be done without use of bending spring.

If a bend is needed near the end of a pipe, form it well back from the end of a long length and saw off the surplus. Connectors and fittings with solvent welded cold-water supply lines are similar to those for use with the solvent weld waste system. Use a fine-toothed hacksaw for cutting all plastic pipes, and glasspaper, file or even pocket knife for removing burrs.

For any work being undertaken in galvanised piping a pair of Stillson wrenches will be required and possibly dies with which to thread the ends of the pipe, which in turn requires a pipe vice. Common steel pipe sizes are shown in the table below.

Pipe bore	Length of thread screwed into fitting at each end
12 mm	12 mm
18 mm	12 mm
25 mm	15 mm
32 mm	15 mm
38 mm	15 mm
50 mm	18 mm

Measuring pipe lengths for cutting must be done accurately and a steel tape is most convenient as it can be used to measure bends. Where these are in the run of the pipe add 50 mm to allow for any loss in the overall length because of the bend being formed. Whatever type of pipe is being used, try to set out the position of connectors or fittings, measure between their end faces and then add the amount to be entered into the fitting at each end; in the case of copper or solvent welded plastic, this is usually the same as the pipe diameter. With push-fit plastic the pipe must be cut a little shorter according to the manufacturer's instructions to allow for expansion (see Chapters 5, 6 and 10).

5 Pipe Fittings and Sundries

In order to fulfil its function a pipe may have to follow a very devious route, making many turns and changes of direction. An array of fittings is available to make this possible. The range is much too numerous to illustrate or describe fully within the limits of this book but all have one thing in common – they are ordered by the size and type of pipe which they will accommodate, e.g. for copper a 15 mm tee will join two 15 mm pipes to a third at 90 degrees; 22 mm elbows, 90 degrees, or obtuse at 45 degrees, will take a 22 mm pipe at each end, whereas a reducing coupling 22 mm x 15 mm will accept a 22 mm pipe at one end and a 15 mm pipe at the other. A limited selection of capillary fittings is illustrated in *Fig. 1.*

An equally extensive range of compression fittings is available for use with either $\frac{1}{2}$H or H grade copper (non-manipulative fittings) (see Chapter 6) or $\frac{1}{2}$H and O grade (manipulative fittings).

Manipulative fittings require the ends of tubes to be flared out and it is common practice to use non-manipulative fittings with adaptors as in *Fig. 2.* Capillary and compression fittings are made to adapt from 'imperial' copper tube to metric sizes (see *Fig. 1.* and Chapter 4).

Adaptor **coupling-slip pattern** Imperial to metric Female Female x copper copper To connect imperial tube to metric tube	**Elbow** copper x copper mm 6 8 10 12 15 22 28 35 42 54 67
Straight tap connector copper x union nut Spigot & fibre washer joint	**Tee** all ends for copper Also available as slip fittings in 15 and 22mm sizes mm 6 8 10 12 15 22 28 35 42 54

Fig. 1.

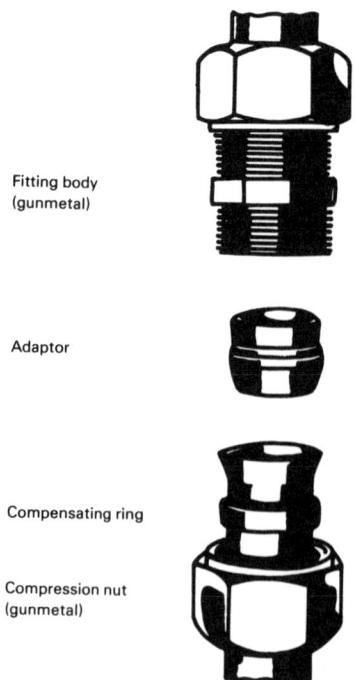

Fitting body
(gunmetal)

Adaptor

Compensating ring

Compression nut
(gunmetal)

Fig. 2.

Re-rounding tool
8mm
10mm
12mm

Re-rounding tool
mm
15
22
28
35
42
54

Tube straightener
6/8mm sizes combined
8/10mm sizes combined

Tube bender
6/8mm sizes combined
10mm
The Yorkshire bender is a hand forming tool designed to put bends in 'Minibore' copper tube

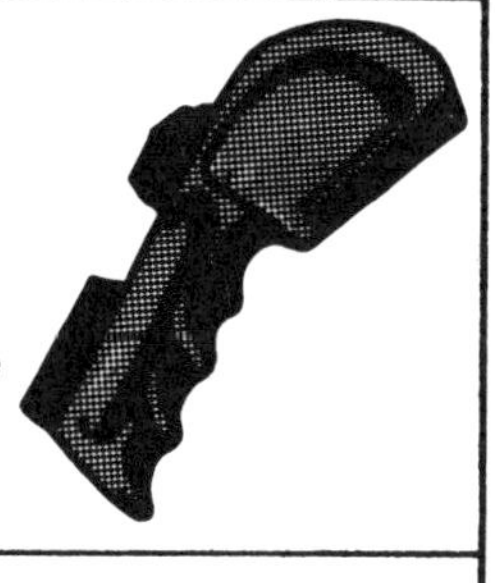

Copper tube forming tool
mm
15
22
28

Fig. 3.

The tools used for flaring are simply punches bevelled at the appropriate angle. Soft copper tube is easily deformed by cutting, but is also equally easily reshaped with a simple re-rounding tool.

Small-bore tube supplied in coils is straightened with a tube straightener. A little lubricating oil is applied and the tool pulled along the tube. An equally simple tool is also used for bending. *(Fig. 3.)*

Both capillary and compression fittings are available threaded male or female to adapt from copper to BSP. These are British Standard Pipe Threads of the type used on plumbing fitments, and galvanised pipework. The threads are often referred to as 'iron' — MI indicating male thread and FI female; *Fig. 4.* illustrates two examples.

mm	MI
12 x	$\frac{3}{8}$"
12 x	$\frac{1}{2}$"
15 x	$\frac{3}{8}$"
15 x	$\frac{1}{2}$"
15 x	$\frac{3}{4}$"
22 x	$\frac{3}{4}$"
22 x	1"
28 x	$\frac{3}{4}$"
28 x	1"
35 x	$1\frac{1}{4}$"
42 x	$1\frac{1}{2}$"

Male elbow
copper x male iron

mm	mm	F1
15 x 15 x	$\frac{3}{8}$"	
15 x 15 x	$\frac{1}{2}$"	
22 x 22 x	$\frac{1}{2}$"	
22 x 22 x	$\frac{3}{4}$"	
28 x 28 x	$\frac{1}{2}$"	
28 x 28 x	$\frac{3}{4}$"	
28 x 28 x	1"	

Female tee, female iron on branch
copper x copper x female iron

Fig. 4.

'Nevastops' are small in-line isolating valves designed for inlet supplies to appliances. They make it possible to service the appliance without turning off the supply elsewhere and being small are not unsightly. They can also be used to restrict the flow rate of water to taps if this is found to be excessive. Four types are available from Deltaflow; as shown in *Fig. 5.*

A similar range of compression fittings to those used for copper is also available for polythene piping. Suitable 'liners', which are colour coded according to the density and working pressure of the piping, must be used (see Chapters 4 and 6); **they are class B-red, C-blue and D-green.**

As stated earlier galvanised pipework is jointed by fittings threaded BSP *(Fig. 6.).* Traditionally these joints were painted and wound with plumber's hemp before being made, but threaded joints in general are now made quite watertight by the use of PTFE plastic tape.

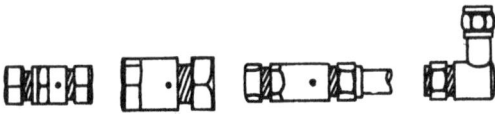

Fig. 5. Nevastops

Fig. 6.

Union Elbows

Cone seat

90° Female equal

Tees

90° Female

Equal & reducing

Unions

Cone-seat, Female equal

Fig. 7.

Saddle band	mm
	6
	8
	10
	12
	15
	22
	28
	35
	42
	54

Single spacing clip	mm
	15
	22
	28

Two-piece spacing clip	mm
	8
	12
	15
	22
	28
	35
	42
	54

Pipe bracket	mm
Unpolished brass strip	15
	22
	28

Fig. 7. indicates a few of the many clips used for supporting runs of pipe, and *Fig. 8.* shows recommended solder and flux.

Solder wire

In tins
Small
(75 grammes)
Medium
(500 grammes)
Large
(3.36 kilogrammes)

Fig. 8.

'Yorkshire' flux
Made specially for 'Yorkshire' fittings

BRITISH STANDARD PIPE THREADS
Dimensions in inches

Nominal bore of tube		App. outside dia. of tube		Outside dia. of thread	Depth of thread	Core diameter	No. of threads per inch
$\frac{1}{8}$		$\frac{13}{32}$		·383	·0230	·337	28
	$\frac{1}{4}$		$\frac{17}{32}$	·518	·0335	·451	19
$\frac{3}{8}$		$\frac{11}{16}$		·656	·0335	·589	19
	$\frac{1}{2}$		$\frac{27}{32}$	·825	·0455	·734	14
$\frac{5}{8}$		$\frac{15}{16}$		·902	·0455	·811	14
	$\frac{3}{4}$		$1\frac{1}{16}$	1·041	·0455	·950	14
$\frac{7}{8}$		$1\frac{7}{32}$		1·189	·0455	1·098	14
	1		$1\frac{11}{32}$	1·309	·0580	1·193	11
$1\frac{1}{4}$		$1\frac{11}{16}$		1·650	·0580	1·534	11
	$1\frac{1}{2}$		$1\frac{29}{32}$	1·882	·0580	1·766	11
$1\frac{3}{4}$		$2\frac{5}{32}$		2·116	·0580	2·000	11
	2		$2\frac{3}{8}$	2·347	·0580	2·231	11
$2\frac{1}{4}$		$2\frac{5}{8}$		2·587	·0580	2·471	11
	$2\frac{1}{2}$		3	2·960	·0580	2·844	11
$2\frac{3}{4}$		$3\frac{1}{4}$		3·210	·0580	3·094	1.1
	3		$3\frac{1}{2}$	3·460	·0580	3·344	11
$3\frac{1}{4}$		$3\frac{3}{4}$		3·700	·0580	3·584	11
	$3\frac{1}{2}$		4	3·950	·0580	3·834	11
$3\frac{3}{4}$		$4\frac{1}{4}$		4·200	·0580	4·084	11
	4		$4\frac{1}{2}$	4·450	·0580	4·334	11
$4\frac{1}{2}$		5		4·950	·0580	4·834	11
	5		$5\frac{1}{2}$	5·450	·0580	5·334	11
$5\frac{1}{2}$		6		5·950	·0580	5·834	11
	6		$6\frac{1}{2}$	6·450	·0580	6·334	11
7		$7\frac{1}{2}$		7·450	·0640	7·322	10
	8		$8\frac{1}{2}$	8·450	·0640	8·322	10
9		$9\frac{1}{2}$		9·450	·0640	9·322	10
	10		$10\frac{1}{2}$	10·450	·0640	10·322	10
11		$11\frac{1}{2}$		11·450	·0800	11·290	8
	12		$12\frac{1}{2}$	12·450	·0800	12·290	8
13		$13\frac{3}{4}$		13·680	·0800	13·520	8
	14		$14\frac{3}{4}$	14·680	·0800	14·520	8
15		$15\frac{3}{4}$		15·680	·0800	15·520	8
	16		$16\frac{3}{4}$	16·680	·0800	16·520	8
17		$17\frac{3}{4}$		17·680	·0800	17·520	8
	18		$18\frac{3}{4}$	18·680	·0800	18·520	8

Angles of thread 55°. Threads rounded at crests and roots leaving depth of thread=0·64 pitch app. Taper screws coned $\frac{1}{16}$ inch (measured on diameter) per inch length.

6 Making Joints in Pipework

Depending upon the age and locality of the property the householder may find his water systems (cold, domestic hot, waste-water and central heating) run in a combination of galvanised steel pipe, lead, copper or plastic. The purpose of this chapter is to explain how joints in these pipes can be made.

It would be unusual now to run interior supplies in steel piping but should this exist, if repairs or extensions are being considered it is advisable to change to copper. Mains supplies may be of steel, and water for agricultural purposes is often run in steel piping as it is tough and durable. However, this too is being superseded by plastic piping which is flexible, less susceptible to bursts and even more economical.

Steel pipework is screwed together using fittings threaded with BSP (British Standard Pipe Thread). Unions may be purchased which are already threaded, but lengths of pipe need to have their ends threaded once they have been sawn to length. This necessitates having a pipe vice and suitable dies and die stock, or ordering the pipe cut to the required lengths and threaded by the merchant. Pipes *in situ* will have rusted and considerable pressure, using a pair of Stillson wrenches, will be required to loosen them. Heating with the blowtorch will help, one Stillson gripping the pipe/or union and the other turning in the opposite direction. The most simple joint is that using a coupling, connecting pipes of the same size which are not expected to be disconnected again in the forseeable future. All the threads should be smeared with jointing compound and those on the pipes then wound tightly with a few strands of plumber's hemp starting about one thread away from the end of the pipe. Grip one pipe with one Stillson, engage the coupling with the pipe thread which is made easier by not having the hemp right to the end, and using the other Stillson screw up tightly. Now change the position of the Stillsons so that the coupling is gripped firmly and screw in the second length of piping. Pull off surplus hemp from the joint and with a cloth smooth off the jointing compound round the coupling. *(Fig. 1.)*

PTFE tape (see Chapter 3) can be used as an alternative to hemp and compound – two thicknesses are bound clockwise round the thread.

Fittings are available for adaption from steel pipe to copper. They are threaded internally (FI) to receive the steel pipe and are hexagonal at that end, so that a spanner can be used to tighten them. Again hemp and compound or PTFE tape should be used to make the joint. The other end will have either capillary or compression fittings to join it to the copper.

The traditional 'wiped' joint is used to connect two sections of lead piping together, or to join brass or copper adaptors or spigots to it, so from that point the system can be run in copper. The end of one piece of the lead piping must be flared slightly; this can be done with a tapered hardwood boss which is hammered into it. The other pipe is then bevelled off with a rasp or knife, so that the two fit snugly together with the bores in line. Paint a ring of plumber's black round each pipe about 30 mm from the end and from this scrape each one bright with a knife or shavehook. The complete joint should be approxi-

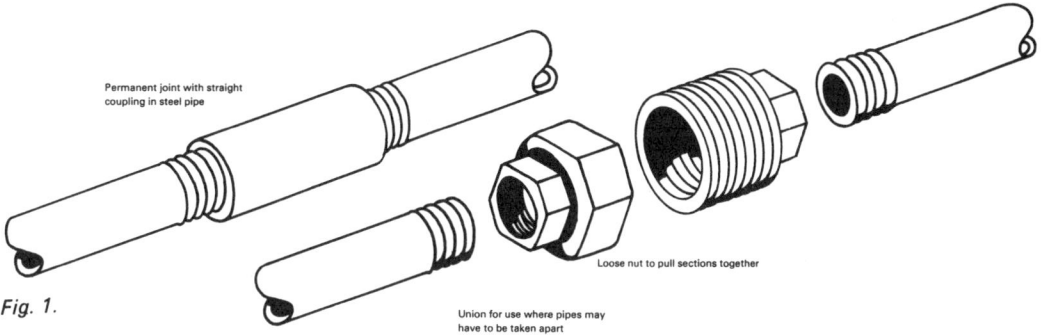

Permanent joint with straight coupling in steel pipe

Loose nut to pull sections together

Fig. 1.

Union for use where pipes may have to be taken apart

mately 75 mm long and the plumber's black restricts the solder between these limits. If adapting to copper, the tail of the adaptor, or alternatively a short piece of pipe to which a capillary or compression fitting can later be added, should be cleaned with glasspaper and tinned (i.e. coated with plumber's solder) — plastic range 183°C to 262°C (70 per cent lead, 30 per cent tin with a little antimony in some qualities).

The tube must be smeared with a noncorrosive flux and the end heated with the blowtorch, so that as it is rubbed with the plumber's solder sufficient melts off to cover it with a thin coating.

The joint must now be held together making sure that each part is in line, if necessary by using some form of clamp, and the whole area of the joint heated evenly, tallow being smeared over it to facilitate the spread of the solder. The stick of solder has to be heated at the same time, so that as it melts solder runs onto the joint. Only apply sufficient heat to keep the solder in a plastic condition, it is essential to pack the rim of the flared end to capacity.

You now require a plumber's moleskin, which is a specially woven cloth pad used for wiping the solder round the joint. It too should be heated and coated with hot tallow to assist the spread of the solder. With the moleskin supported by all four fingers, the semi-molten solder is spread firmly over the joint. Adjusting the pressure, tapers it off towards the ends. Apply further solder as required taking

care that the joint is not allowed to cool until it is finally complete. If necessary keep the solder workable by the application of further heat. Once sufficient solder has been applied to produce the typical rounded mass over the joint, it should be given a final wipe round to produce a smooth finish. *(Fig. 2.)*

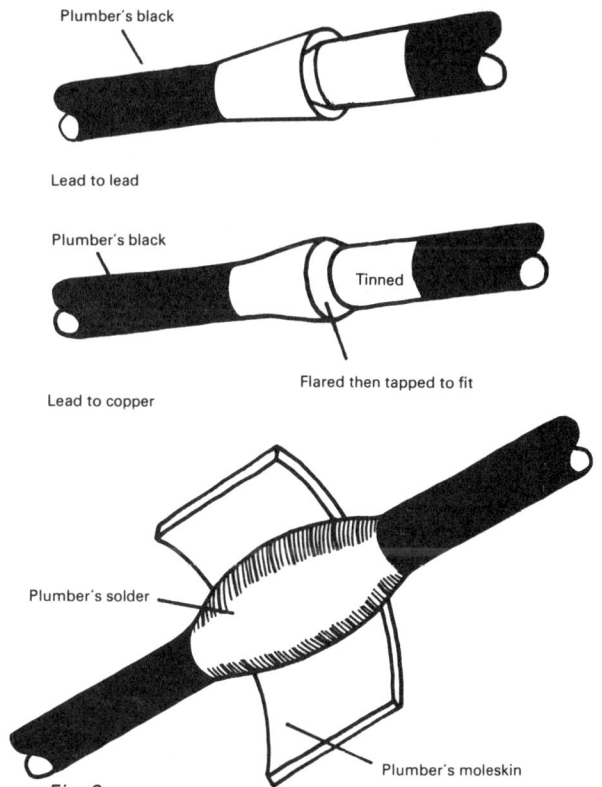

Plumber's black

Lead to lead

Plumber's black

Tinned

Flared then tapped to fit

Lead to copper

Plumber's solder

Plumber's moleskin

Fig. 2.

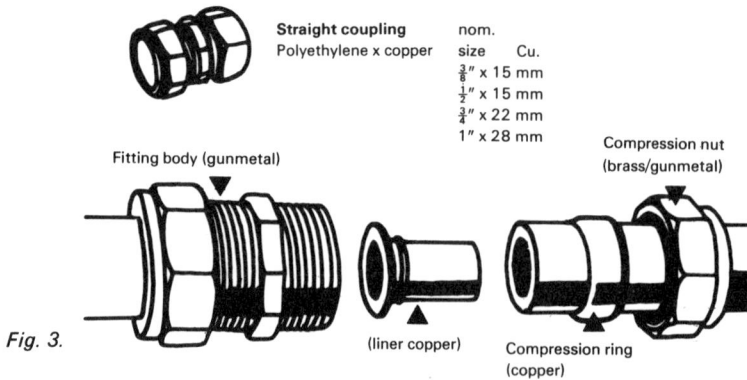

Straight coupling
Polyethylene x copper

nom.	
size	Cu.
$\frac{3}{8}$"	x 15 mm
$\frac{1}{2}$"	x 15 mm
$\frac{3}{4}$"	x 22 mm
1"	x 28 mm

Fitting body (gunmetal)

Compression nut
(brass/gunmetal)

Fig. 3.

(liner copper)

Compression ring
(copper)

If sheet metal is tinned, it can be 'sweated' together. The two pieces should be cleaned thoroughly, if necessary by scraping or filing, then glasspapered and fluxed. The purpose of the flux is to prevent oxidisation of the metal as it is heated and to assist the flow of the solder. Solder is then applied either by rubbing it on to the heated metal or with a soldering iron. Both are again fluxed, and brought together under heat until both layers of solder fuse together. In sheet metal-working this has many applications and is the method suggested for the repair to the ball valve in Chapter 7.

It should be remembered that pipes full of water are heavy and that lead, copper and plastic will sag if not properly supported by the appropriate fittings as recommended by the manufacturer (see Chapter 10).

Compression fittings for plastic piping are similar to those for use with copper, but a flanged inset must be placed in the pipe to prevent distortion of the flexible plastic. If necessary the end of the pipe can be softened by gentle heating, so that the insert can be fitted easily. *(Fig. 3.)*

The recommended procedure is as follows:

1. Use correct type, class and size of polythene pipe for the situation (see Chapter 4).
2. Make sure pipe is round, ends squared and burrs removed.
3. Slip compression nut and compression ring over pipe.
4. Insert correct liner (see Chapter 5) into pipe until flange touches pipe end.
5. Push pipe end into body of fitting up to the shoulder or stop in the bore.
6. Hand-tighten compression nut.
7. Make a further one and a half turns with a spanner to complete the joint from the point at which the compression ring begins to grip the polythene pipe. An arrow head on the coupling nut is a useful guide. When correctly tightened the pipe cannot be twisted within the fitting.
8. For large fittings in particular, remove all traces of dirt or grit from the thread of the body and nut, and apply a few drops of light oil to facilitate tightening. Spanners of at least 600 mm length should be used because of the amount of torque necessary to tighten the compression nut.

Capillary fittings make the neatest joints in copper tube *(Fig. 4.)*, but compression joints are simpler. For either type the ends of the tube must be cut square with a hacksaw or pipecutter. A piece of masking tape wrapped round the tube provides an edge to cut to, ensuring squareness. Pipe cutters are rather like G-clamps and are fitted to the tube and rotated, the pressure being gradually increased until the cutting wheel severs the tube. Both methods raise burrs which must be removed from inside and outside the pipe by filing. If end-feed capillary fittings are being used,

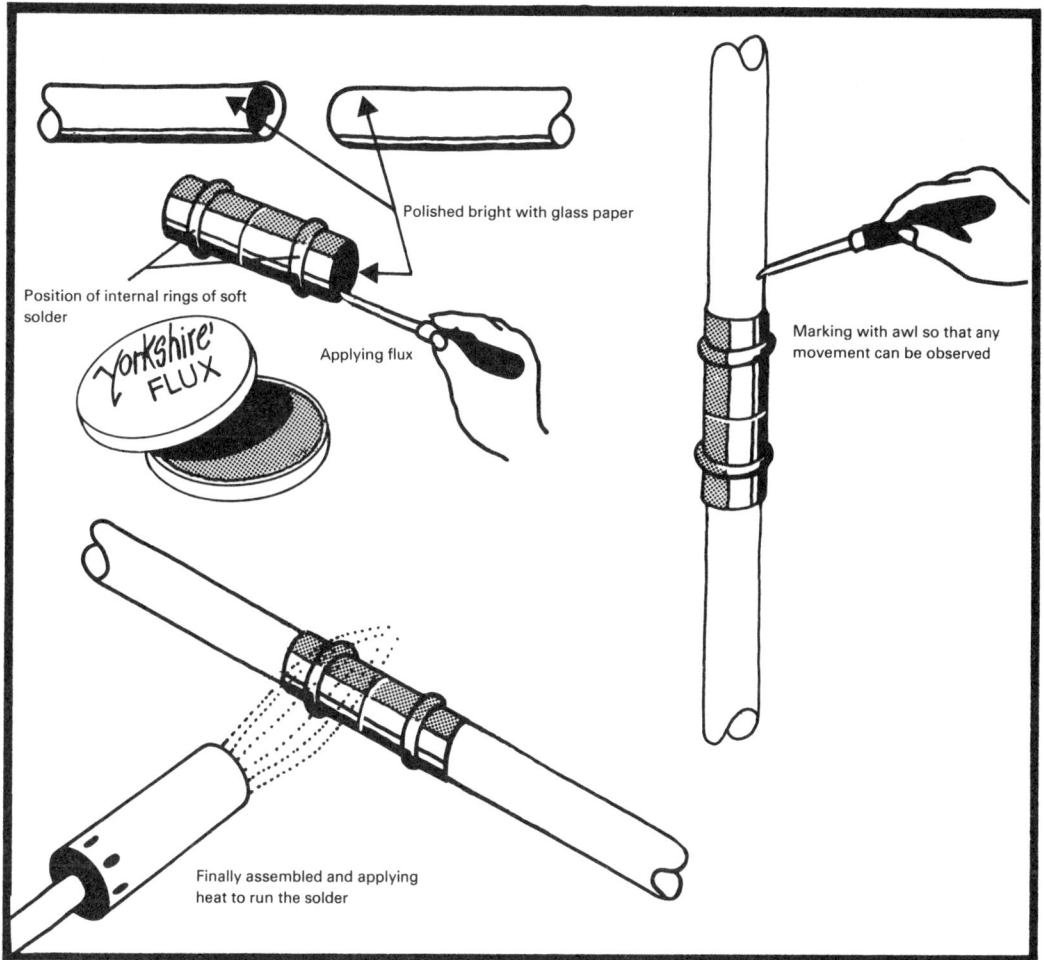

Polished bright with glass paper

Position of internal rings of soft solder

'Yorkshire' FLUX

Applying flux

Marking with awl so that any movement can be observed

Finally assembled and applying heat to run the solder

Fig. 4.

solder must be applied at the fitting mouth as the joint is being made.

Presoldered joints have a sunken ring of solder inside and are much easier to use. The inside of the fitting and the ends of the tube must be cleaned with fine glasspaper and pressed together up to the stop inside the fitting, mark the length of pipe entered with a scratch. The pipes are then withdrawn from the union, a non-corrosive flux applied to each and the joint reassembled. The scratch will indicate correct fitting and heat is then applied by a gas blowlamp to the union and the pipe. Heating should be overall and the solder

will melt and creep between the tube and union by capillary attraction until a bright ring of solder appears all round the joint. A ceramic tile behind the joint will reduce the fire risk. When cool, surplus flux should be wiped off. When necessary, joints may be taken apart by reheating, and then withdrawing the tube. It can be remade by fluxing the tinned end and reheating until it can be pushed fully home once more, adding extra solder if required. New pieces of tube must be tinned before use in such a situation.

Compression joints do not rely upon solder, so no heat is required. To assemble,

Nut

Cones

Nut

Fig. 5.

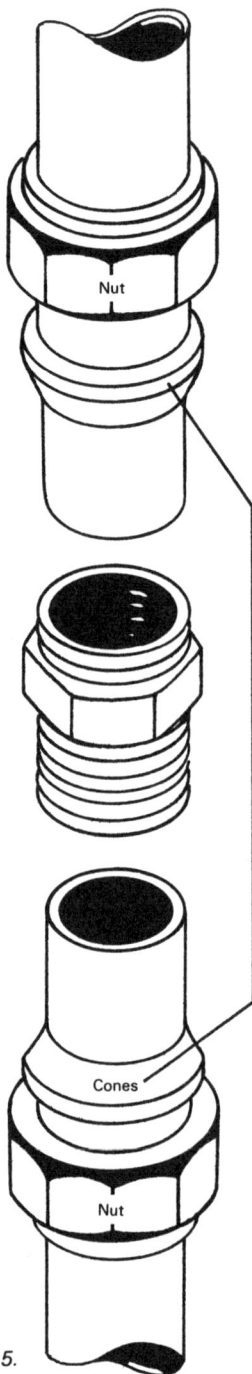

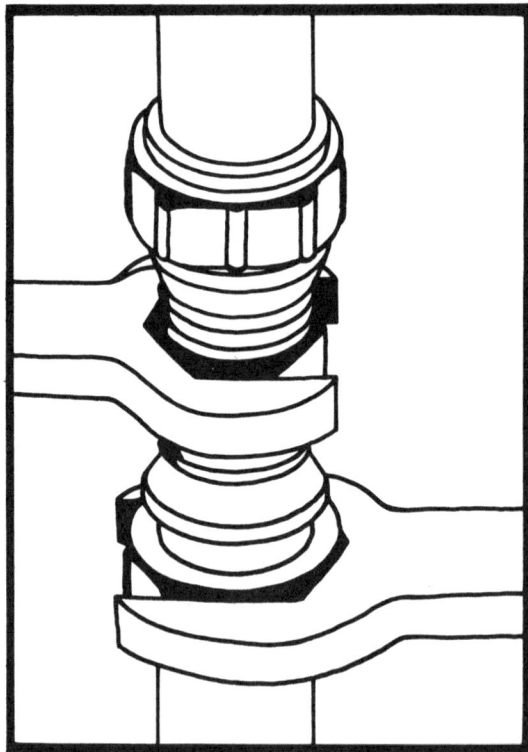

Assembling using two spanners

slide the nut onto each pipe followed by the cone (other names are olive or compression ring). If this has a long chamfer, this should be towards the end of the pipe. The pipes should then be pressed in up to the stop and the union assembled finger-tight, then the pipes marked as before, so that any tendency to creep out, as the nuts are tightened with spanners, can be observed and checked. A second spanner should hold the fitting as each nut is tightened, three turns are usually sufficient. Overtightening may result in damage to the threads or cone resulting in an unsatisfactory joint. *(Fig. 5)* Manufacturers do not recommend the use of a jointing compound preferring the metal to metal connection of the accurately machined fitting.

Three types of joints are also employed for plastic pipework. They are by compression nut, solvent cement welding or push-fit joints sealed by rubber 'O' rings located in a moulded housing and retained by a snap-in clip as in the Bartol System. *(Fig. 6.).* Cement welding may be used for both cold-water supply systems and waste-water pipes, whereas the push-fit system is for waste water only, as this is not under pressure. To make the cement welded joints:

1. Ensure that the pipe end is cut square and remove burrs with a coarse file. Apply solvent cleaner to both areas to be joined.
2. Apply Bartol solvent cement to both clean surfaces to be joined.
3. Insert the pipe into the socket with a slight twist. Clean off surplus cement. The joint can be handled with care in less than five minutes. Wash out the

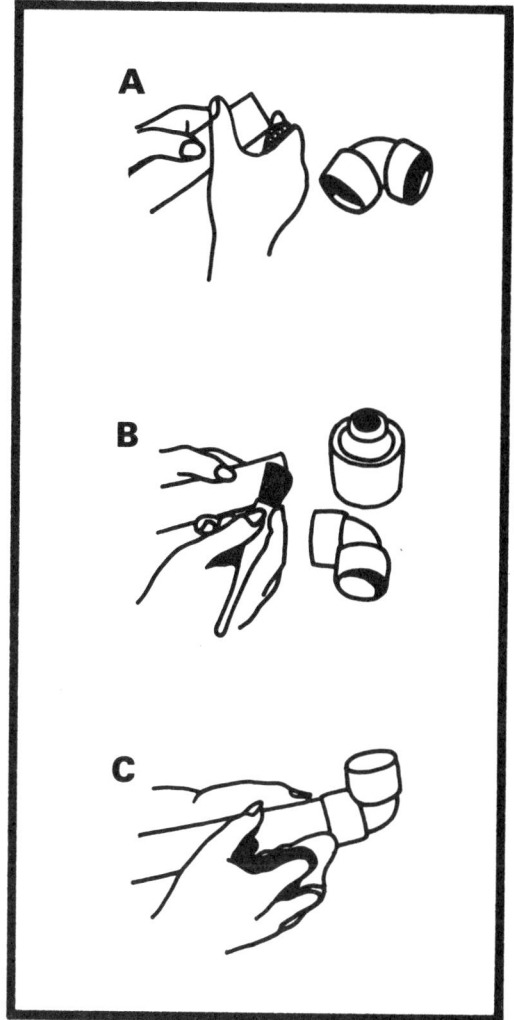

Fig. 6.

brush in a cleaner.

The jointing sequence for push-fit pipework is illustrated in *Fig. 7*. The parts are lubricated with soft soap before assembling.

JOINTING SEQUENCE

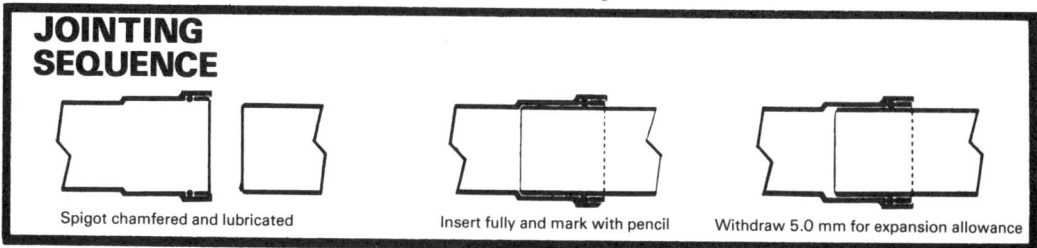

Spigot chamfered and lubricated

Insert fully and mark with pencil

Withdraw 5.0 mm for expansion allowance

Fig. 7.

7 Renewing Washers

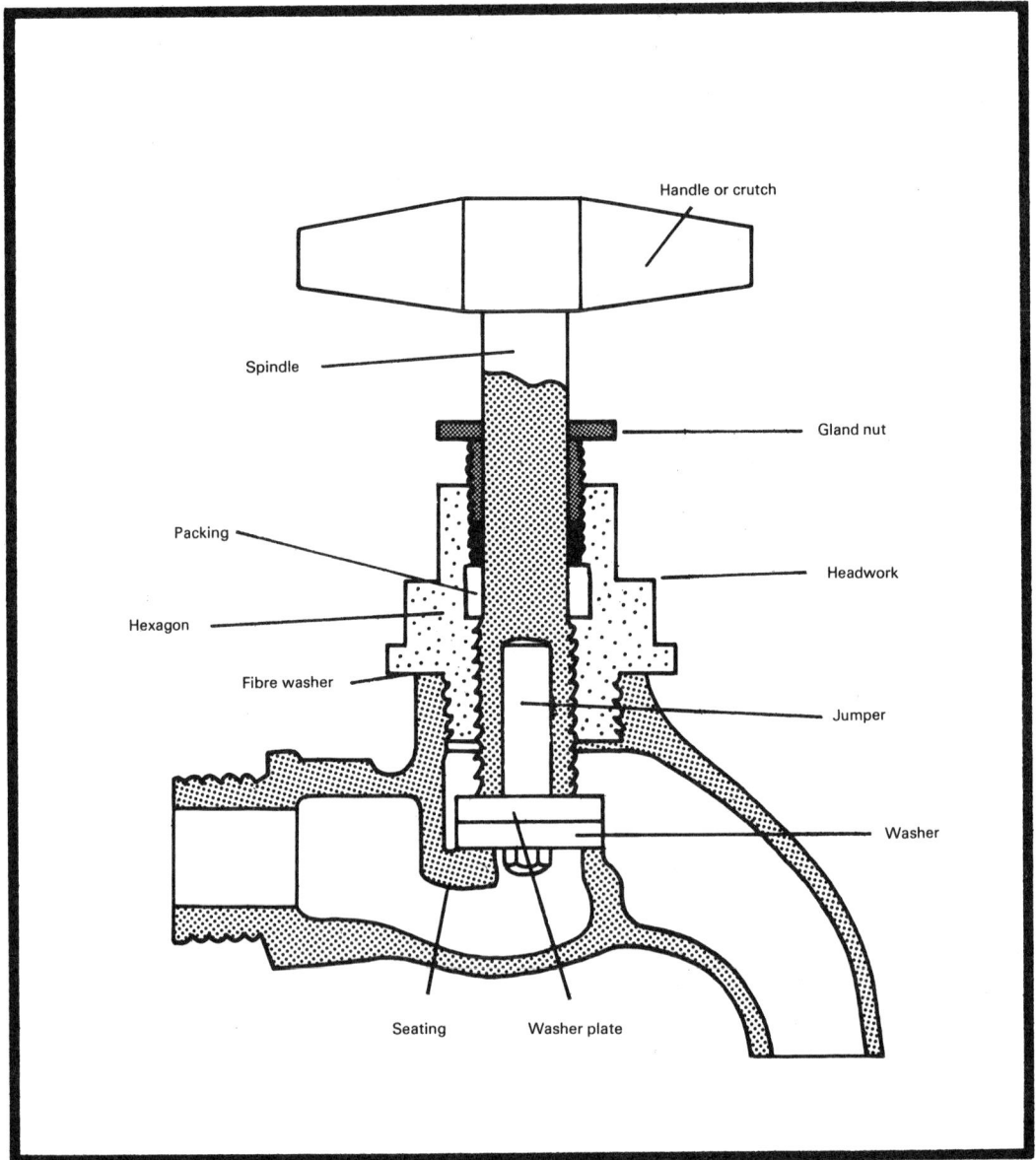

Fig. 1. Section through a tap

Water should cease to flow or drip from a tap which is closed without undue turning force having to be applied to the crutch or handwheel. As these are rotated the spindle is turned applying pressure to, or relieving pressure on, the washer. A mechanical device such as a tap is necessary to regulate the flow of water from the domestic system because of the water pressure. *(Fig. 1.)* Water is supplied by the authorities to the consumer at pressures of between 30 lb/sq in and 80 lb/sq in, in accordance with the 1945 Water Act. These will raise the water from a minimum height of 21·329m (70ft) to a maximum of 54·846m (180ft) respectively, depending upon requirements; to convert to metric pressures — 14·05lb/sq in = 1 bar. In older direct systems this is the pressure at each cold tap or ball valve. The more modern indirect system (see Chapter 2) results in a lower pressure at both cold and hot taps. By this method, both are gravity fed and the pressure depends upon the height of the cold-water cistern above the tap. The greater the difference in levels the bigger the pressure and flow rate.

The tap washer is located on the underside of the jumper or washer plate and is held in place by a small nut. *(Fig. 2.)* The spigot of the jumper fits into a hole in the bottom of the spindle, which in some taps is held in and in others is quite loose. In either case it is a sliding fit being free to turn. Therefore as the spindle presses the washer onto its seating it stops turning, thus preventing wear. In stop taps, and cold-water taps in direct systems, the jumper is quite loose, so should there be a drop in mains pressure when the tap is open, the washer will drop on to its seating preventing a back-syphonage of water into the mains and causing possible contamination. Taps used on indirect systems are designed to lift the washer from its seating as the tap is opened because of the lower water pressure (see Chapter 2). If a tap drips after being closed, or makes odd noises when open it is an indication

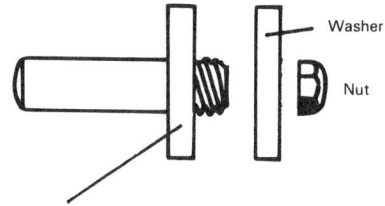

Fig. 2.

that the washer should be renewed. A washbasin or bath may be badly stained by water which is constantly dripping. The author can remember being called to advise on such a stain in a bath. A considerable area below the tap was green, obviously arising from the copper pipework and a solvent in the water. It was not until the washer was replaced, and after several weeks of regular application of a cleaner, that the stain was finally removed. It should also be remembered that a dripping tap is a potential source of trouble in frosty weather if the stackpipe is external as the trickle of water will freeze rapidly in the hopper head. In these days of energy conservation it should be remembered that if it is a hot-water tap, fuel is also being wasted. A few years ago it was customary to fit hot taps with red fibre or hard rubber washers, and cold-water taps with leather or soft rubber. It is now common practice to use synthetic rubber washers which are suitable for both hot and cold water. Be sure to have the correct diameter washer available for the size of tap. Tap sizes are measured by the bore of the pipe to which they are fitted.

The steps taken to renew the washer are as follows:
1. For cold taps on a direct supply, turn off at the consumer stop tap. Where the system is indirect, turn off at the consumer stop tap if the faulty tap is at the sink, or at the appropriate wheel valve nearest to the tap. Such wheel valves are often situated in the hot-water cylinder cupboard, but if they have not been provided, the water may be cut off at the cold-water cistern.

This is done with string, tying the arm of the ball valve tightly to a piece of wood placed across the edges of the cistern to keep the valve closed as water is drained off. *(Fig. 3.)*

2. When the necessary steps have been taken, the faulty tap should be opened and all water allowed to drain from it.

3. Unscrew the bell-shaped shield which is designed for easy cleaning. (Stop taps and those designed for outdoor use do not have these fitted.) The shield should only be finger-tight, but some do have flats around their bottom edge to which an adjustable spanner can be applied. Failing this, if a mole wrench, or Stillson, has to be used, protect the chromium plating with a rag, or heavy duty adhesive tape, but a strap wrench (see *Fig. 11.*, Chapter 3) is the ideal tool.

4. When the cover is lifted up to the underside of the crutch, a large nut is exposed, which is part of the spindle casing or tap headgear. This must be unscrewed using a plain spanner of a suitable size across the flats, or an adjustable spanner. While this is being done, the tap should be supported firmly by hand, or with a suitably padded wrench to avoid undue strain

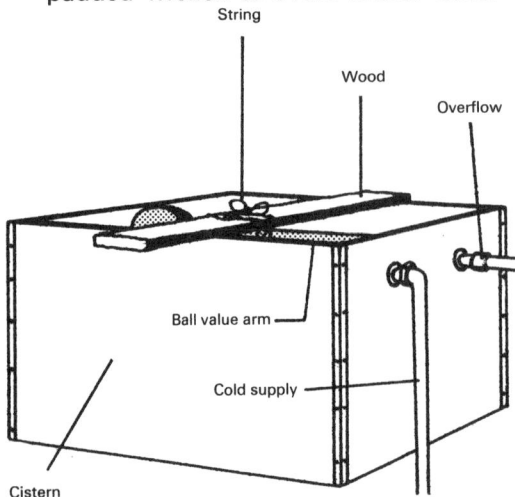

Fig. 3. Tying up ball valve arm

being put on it, or its connection with the washbasin, bath or pipe. If, because of age, this casing is difficult to unscrew, a smart blow with a hammer at the far end of the spanner will usually free it.

5. Once unscrewed, this spindle assembly, or headgear, may now be lifted clear of the tap.

6. Lift out the jumper from the body of the tap, if it is of the loose variety, and unscrew the small retaining nut on the underside. If a fixed jumper, it will have been lifted out with the head. The spigot or the washer plate of the jumper may have to be held in a wrench or with pliers, while a small spanner is applied to the nut.

7. Twist off the old washer and fit the new one.

8. Now replace the retaining nut and tighten up. If it bites a little way into the washer, this helps to lock the nut.

9. You will have found a thin red fibre washer between the spindle assembly and the body of the tap. If this happens to be damaged it should be renewed, but in the author's experience these washers are not always easily obtained, so care is advisable.

10. Examine the washer seating within the body of the tap to ensure that it is not damaged. Abrasions may be seen, or felt with the finger. If there is damage to the seating this should be made good with a reseating tool (see *Fig. 34.*, Chapter 4). This is the simplest pattern. More complex types are available for use on both tap and ball valves. Your local tool hire firm should be able to help with this.

11. Replace the jumper in the spindle if it was loose.

12. Screw in the assembled head complete with fibre washer and tighten down with a spanner. As previously, support the body of the tap to avoid strain.

13. By hand, screw down the shield.

14. Close the tap.

15. Turn on the supply once more and/or release the ball valve.
16. Open the tap slowly so that air is expelled from the system, then check that the tap is working efficiently.

One modern version of the pillar tap (see Chapter 9) has its nozzle and hand-wheel combined to point down at the sink. With this pattern there is no need to turn off the supply when renewing a washer. As the head is removed a valve in the tap body closes cutting off the water.

Stop taps are heavy duty and are designed to give exceptionally good service over many years, seldom giving much trouble. Consumer stop taps can be re-washered in the same way as any other tap if they fail to shut off the supply, providing there is a company stop tap by which they can be isolated. Be prepared for water draining back out of the system once the tap headgear is removed.

If the water authority's stop tap proves to be inefficient then the authority should be notified.

Ball Valves

A ball valve *(Fig. 4.)* will control the flow of water into your cold-water cistern or header tanks for the hot-water system or central heating. It is also the common method controlling the supply to WC cisterns. The level of water is governed by the rise and fall of the floating ball. When water is drawn off, the ball is lowered and the valve allows water to enter the tank. As the cistern refills the water is cut off once more as the ball rises. Should the cistern overfill, water will drip or trickle from the overflow pipe, which extends to the outside of the building. Occasionally ball valves will stick causing this to happen, but it is usually an indication that the washers should be renewed. It is

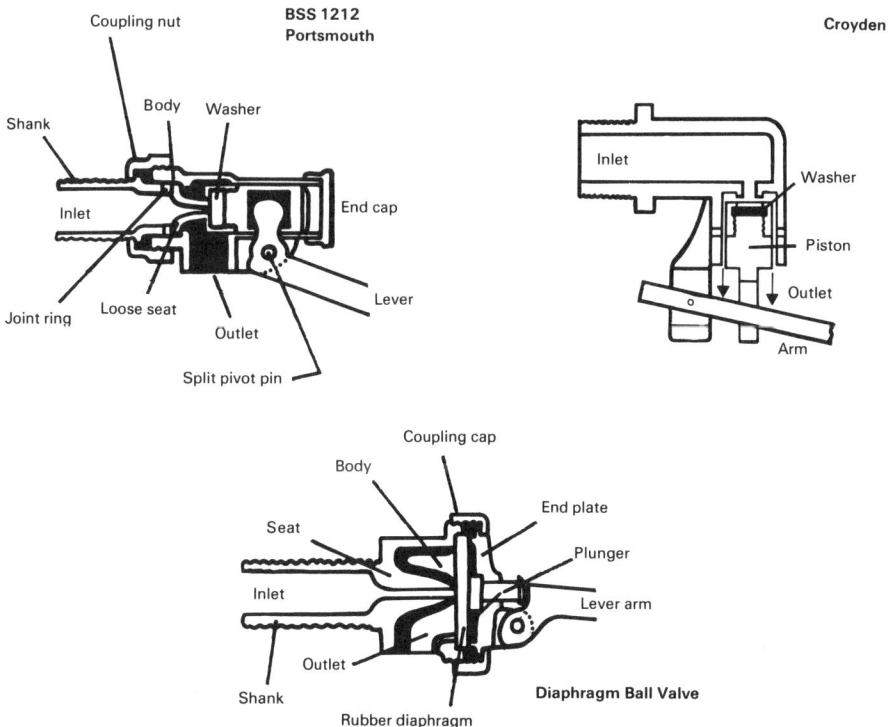

Fig. 4.

just possible, however, that the lever arm connecting the ball and valve is slightly bent upwards so that the water is not cut off soon enough. If you suspect that this is the case, bend the arm carefully downwards, then drain off some of the water, checking that the water level stops about 25mm below the overflow when it refills. The cistern may also overfill if the ball leaks, allowing it to become heavy with water. Should this be the case, tie up the lever arm as indicated earlier, and unscrew the ball from the end. If it is made of copper it may be possible to drain out the water, warm it gently to dry it thoroughly, clean with glasspaper, and sweat on a thin copper patch with soft solder (see Chapter 6). Failing this, or if the ball is made of plastic, it is wiser to renew it completely. The ball valve will be high pressure (HP) or low pressure (LP) depending on whether or not it is part of a direct or indirect system. HP or LP will be stamped on it, and should it be necessary to replace it completely be sure to obtain the correct type.

To replace the valve washer follow this procedure:

1. Turn off the supply at the stop tap and partially drain off the cistern.
2. Slip a piece of string through the eye of the split pin and secure to a convenient point. This is to avoid losing the pin in the cistern if it is dropped. Nip the open ends together with pliers and withdraw the pin.
3. Lift out the ball and arm on to an old cloth or newspaper as it will be wet, and withdraw the piston.
4. Place a screwdriver in the piston slot to give leverage and with combination pliers or a mole wrench unscrew the piston end.
5. Fit the new washer to the end and screw back. Smooth off any burrs which may have been raised.
6. Replace the piston with its slot in the correct position to receive the arm once more and replace the split pin. Open the ends with a screwdriver just sufficiently enough to prevent it from withdrawing accidentally.
7. Test that the arm and piston move freely by raising and lowering the arm manually, then turn on the supply again at the stop tap.
8. Make a final check that the water level is 25mm below the overflow when the valve cuts off. The incoming water from the valve may be directed downwards into the cistern by a short plastic tube which restricts noise. Check that this is screwed firmly into its socket.

Corrosion and residual dirt in cisterns are sometimes a problem — this is dealt with in Chapter 16.

8 Repairing Leaking Glands in Taps

The reader is referred to *Fig. 1.* in Chapter 7, which shows the section through a tap illustrating the usual arrangement of the various parts. It will be appreciated that as the tap is opened the water is directed upwards around the screw threads at the base of the spindle. This is because of the pressure of the water which causes considerable turbulence around the washer before the water is controlled and directed outwards by the nozzle of the tap. The more often a tap is used the more wear there is on the screw threads around the spindle and the greater the tendency for water to escape upwards. Any which does so is contained by the packing and the gland nut. The possibility of water being driven upwards in this way is increased if, by the attachment of some other domestic equipment, the flow from the nozzle is restricted. Water escaping through the packing will ooze out above the gland nut and trickle down the headgear. If an easy-clean shield is fitted this will gradually fill with water and after a time there will be an accumulation of dirt within it which becomes unhygienic, and which in turn leaks out around the spindle.

In older taps the packing consists of graphite impregnated string, but in newer ones a plastic sleeve may be used. Constant use of the tap results in the packing becoming less tightly fitted around the spindle, but by tightening down the gland nut it can be further compressed so taking up the wear. This is easily done on those taps not fitted with a shield so that the headgear is exposed. However, if a shield is fitted the crutch must be removed (or the handwheel) to give access to the gland nut. The crutch fits onto a square section at the top of the spindle, and by the removal of a small screw it can be lifted, or gently tapped off. Take care not to lose this small screw down the waste pipe — put the plug in place! Once the crutch has been taken off, lift the shield and tighten down the gland nut as indicated previously.

Should this prove to be ineffective the packing should be renewed completely, the easiest way generally being to cut off the water and strip down the headgear, so that the old packing can be easily removed from the 'stuffing box'. The new impregnated string, or soft white string charged with tallow, must be compressed into the stuffing box with the spindle in place, and the gland nut then tightened down. The rotation of the spindle should now be tested by replacing the headgear into the body of the tap and turning with the crutch. The packing should fit firmly round the spindle under the pressure of the gland nut, and at the same time it should be possible to rotate the spindle without undue force being necessary. It will be possible to feel the friction between the spindle and the packing and assess its efficiency.

This can then be tested by turning on the water and if necessary making further adjustments to the gland nut. Once this is found to be satisfactory, any remaining dirt should be cleared from the headgear. Finally, screw down the shield finger-tight and refit the crutch.

Obviously it is sensible to check the packing at the same time as a washer is being renewed, or vice versa.

9 Fitting Taps, Modern Tap Tops, Mixer Taps

Until quite recently taps were manufactured to BS 1010 which for twenty-five years or so specified the dimensions to which they should be made for use with sinks, washbasins and baths. Their overall sizes were dependent upon whether they were for use with $\frac{1}{2}$in bore pipes (sinks and washbasins) or $\frac{3}{4}$in bore (baths).

Three common types were made *(Fig. 1.)*, many of which are still in use.

a. Bib tap   – kitchen use with sinks.
b. Pillar tap – standard fitting for washbasins and baths.
c. Globe tap – fitted to baths depending upon their design.

Maintenance of these taps is outlined in Chapters 7 and 8.

The new British Standard BS 5412 represents a fundamental change in permitted water fittings being concerned with the performance in use, adequate flow rates, reliability and durability. This has allowed manufacturers to develop individual designs. One such design is illustrated in *Fig. 2.*, its main features are:

■ A lubricated spindle chamber from which water is excluded.
■ Closing friction which is eliminated by a PTFE thrust washer.
■ Washer arrangement and seating which reduces noise in the supply pipes.
■ Reduction of tap water washer wear.
■ Elimination of weeping glands.
■ General ease of operation.

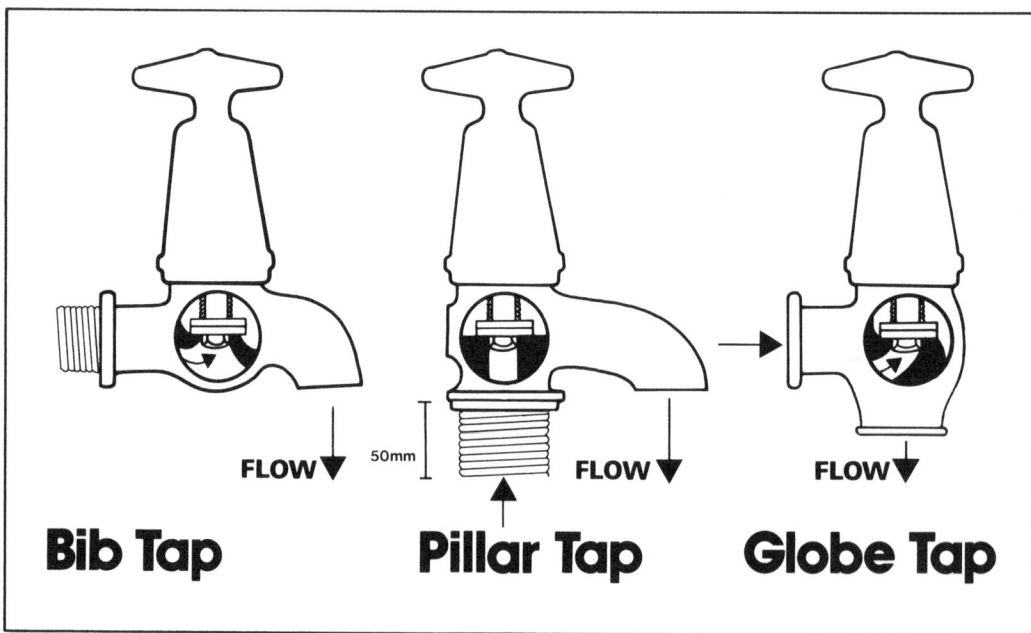

FLOW ▼   50mm   FLOW ▼   FLOW ▼

Bib Tap   **Pillar Tap**   **Globe Tap**

Fig. 1.

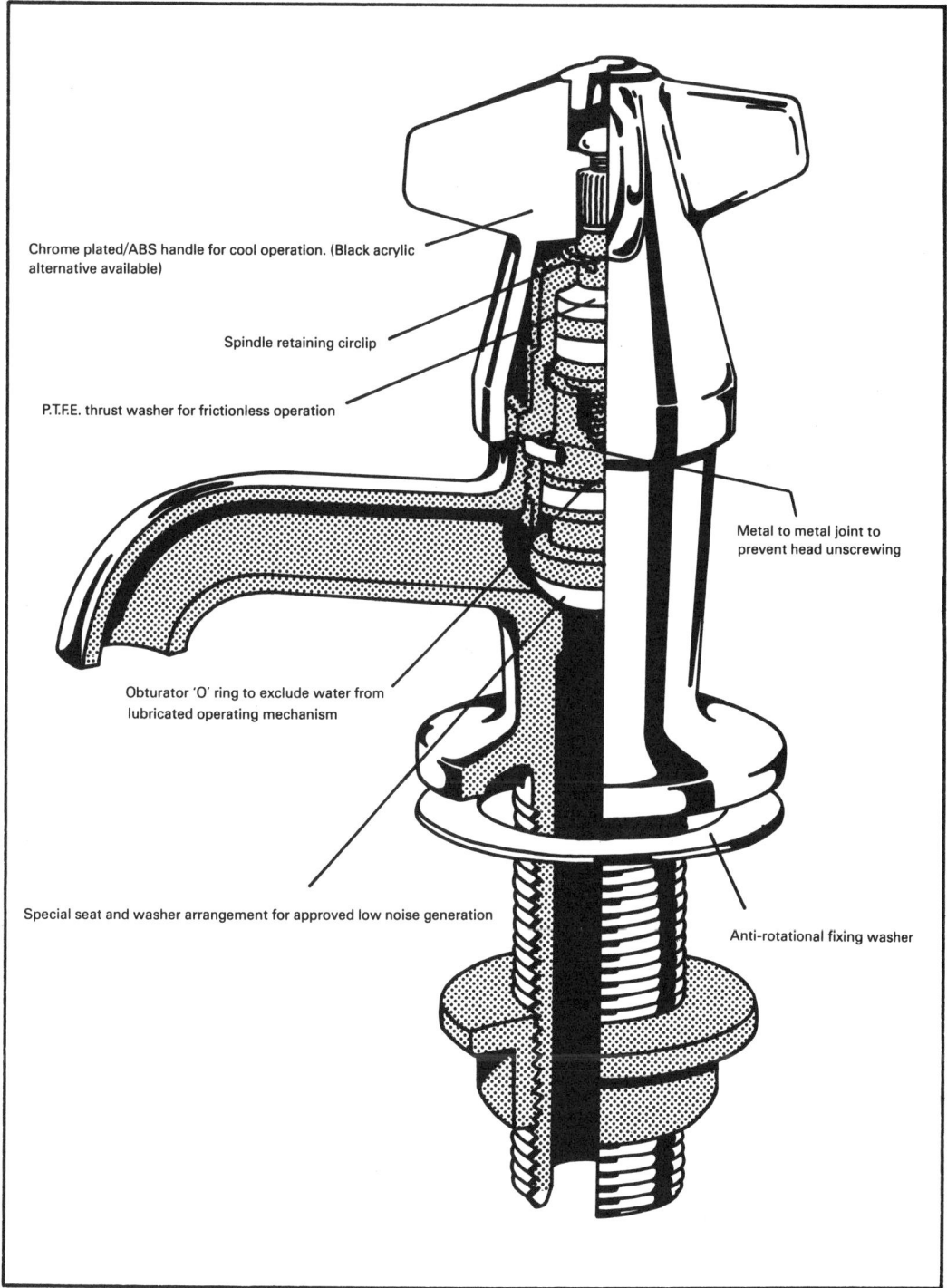

Chrome plated/ABS handle for cool operation. (Black acrylic alternative available)

Spindle retaining circlip

P.T.F.E. thrust washer for frictionless operation

Metal to metal joint to prevent head unscrewing

Obturator 'O' ring to exclude water from lubricated operating mechanism

Special seat and washer arrangement for approved low noise generation

Anti-rotational fixing washer

Fig. 2.

The new British Standard has enabled manufacturers to demonstrate that taps do not have to be ordinary to be functional or because they must be economical they will be inelegant. Modern taps are meticulously engineered and attractively designed. *Fig. 3.* indicates the range of taps available in one such design.

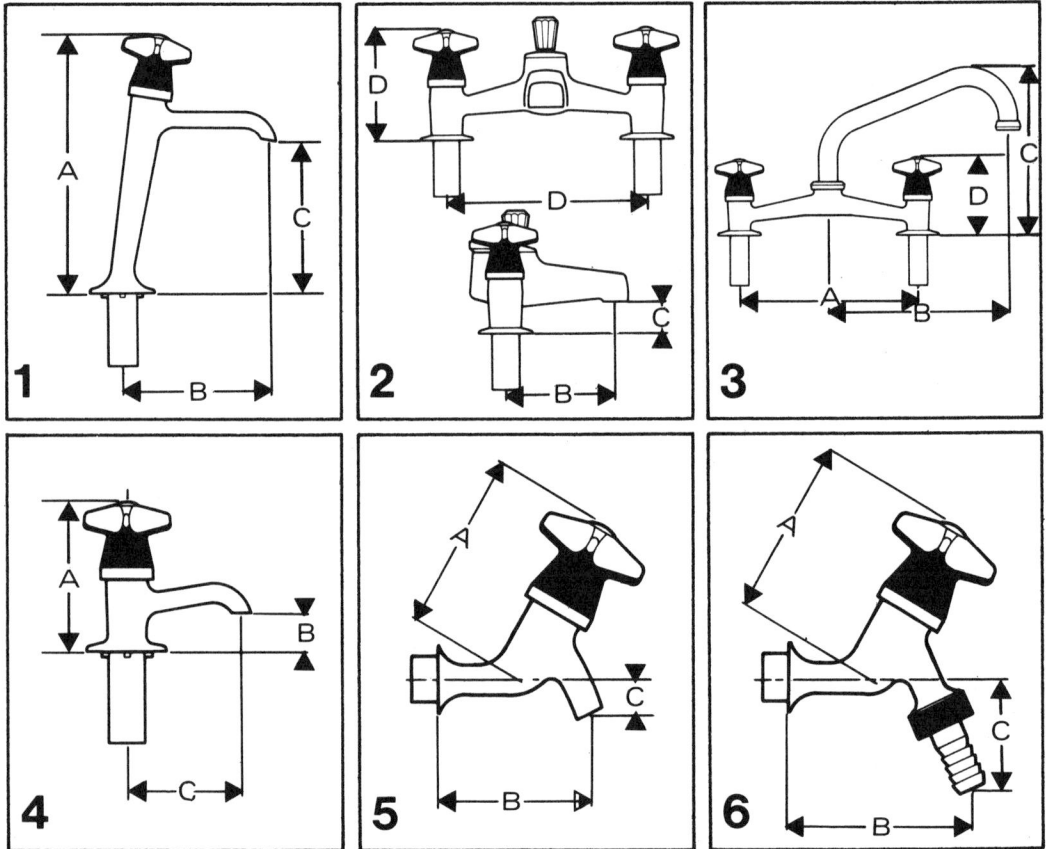

1   **Inclined sink tap**
    ½″ BSP sink pillar
    A: 172 mm
    B: 100 mm
    C: 100 mm

2   **Pillar bath mixer**
    ¾″ BSP. Complete with diverter. Suitable for exposed shower units.
    A: 180 mm
    B: 100 mm
    C: 26 mm
    D: 93 mm

3   **Dual flow sink combination unit**
    ½″ BSP pillar fitting.
    Dual flow outlet
    A: 180 mm
    B: 180 mm
    C: 163 mm
    D: 75 mm

4   **Bath tap**
    ¾″ BSP bath pillar
    A: 92 mm
    B: 82 mm
    C: 25 mm

5   **Bibcock**
    ½″ BSP bicock
    A: 73 mm
    B: 66 mm
    C: 15 mm

6   **Bibcock with hose union**
    ½″ BSP bicock fitted with hose union
    A: 73mm
    B: 79 mm
    C: 43 mm

Fig. 3.

Existing taps can be given a modern styling by exchanging the crutch and shield with a new handle and 'headwork assembly'. *(Fig. 4.)* Kits are available for $\frac{1}{2}$in and $\frac{3}{4}$in taps, the method being as follows:

1. Shut off the hot and cold water supplies, open the taps and leave to drain.
2. Expose the hexagon. If your tap has a crutch or capstan handle with easy-clean shield, then open the tap fully. Unscrew the shield and hold it as high as possible under the handle, exposing the hexagon.
3. Put a spanner on the hexagon and turn anti-clockwise until the headwork assembly detaches. When removing old headworks jerk treatment is more effective than a strong pull. Support the body of the tap (c) to prevent damage to the bath or basin. Before fitting your new top, take a look inside the tap body. Clean out any scale or solids, but in particular, inspect the seat (d). It should be perfectly smooth. If it is in any way pitted or ridged, it can be recut, using a standard $\frac{1}{2}$in or $\frac{3}{4}$in BSS 1010 reseating tool, which can be hired. Alternatively, a qualified plumber can do the job quickly and inexpensively. This is important because a bad seating will cause a continuous leak. You are now ready to insert your new top.
4. Screw the spindle of the new top into the fully open position, so that the washer is raised.
5. Separate the handle from the headwork assembly simply by lifting out the press-fit indice (e) and undoing the retaining screw (f).
6. Fit the head washer (g). Then fit the headwork into the tap body, tightening the hexagon (b) with your spanner.
7. Finally, push the domed handle (a) firmly on to the spindle, tighten the retaining screw, insert the indice, close the tap and turn on the water supply.

Rather than give your taps a facelift you may decide to replace them, in which case here are a few practical points. As previously, turn off the water, drain the taps and disconnect the union with the pipe. Support the tap at all times to avoid strain on the fitment, particularly if it is of vitreous china as are most washbasins. Undo the backnut with the basin wrench and carefully withdraw the whole tap which, if bedded in putty, may be quite solid. Old putty may have to be picked out from behind or below. Once the old tap is released clean up the fitment thoroughly, then proceed as follows:

1. The tap must discharge cleanly into the fitment. Check dimensions.
2. Be sure that indices are correctly positioned – H to hot and C to cold.
3. Fit the taps symmetrically using the anti-rotational washer.
4. It is no longer recommended practice to set the tap in a hard-setting compound such as putty – use a modern non-setting type.
5. Secure with backnut, tightening down onto the thick plastic washer which is now normally supplied with a new tap.
6. Remake the union with the supply pipe, smearing it and the threads with pipe-jointing compound, or use PTFE tape. Formerly taps had 'tails' $2\frac{1}{2}$in long. The new British Standard is 50 mm, so if the supply pipe is rigid you may need an adaptor to make up the difference. These are available for $\frac{1}{2}$in and $\frac{3}{4}$in taps.
7. Turn on the supply – check for leaks. Note that the tap should shut off with light pressure only.

After use if hard screwing down is necessary it is an indication that the washer should be replaced.

Many modern styles of tap now incorporate a handwheel moulded in acrylic, the advantages being both a sparkling appearance and, as it is a good insulator,

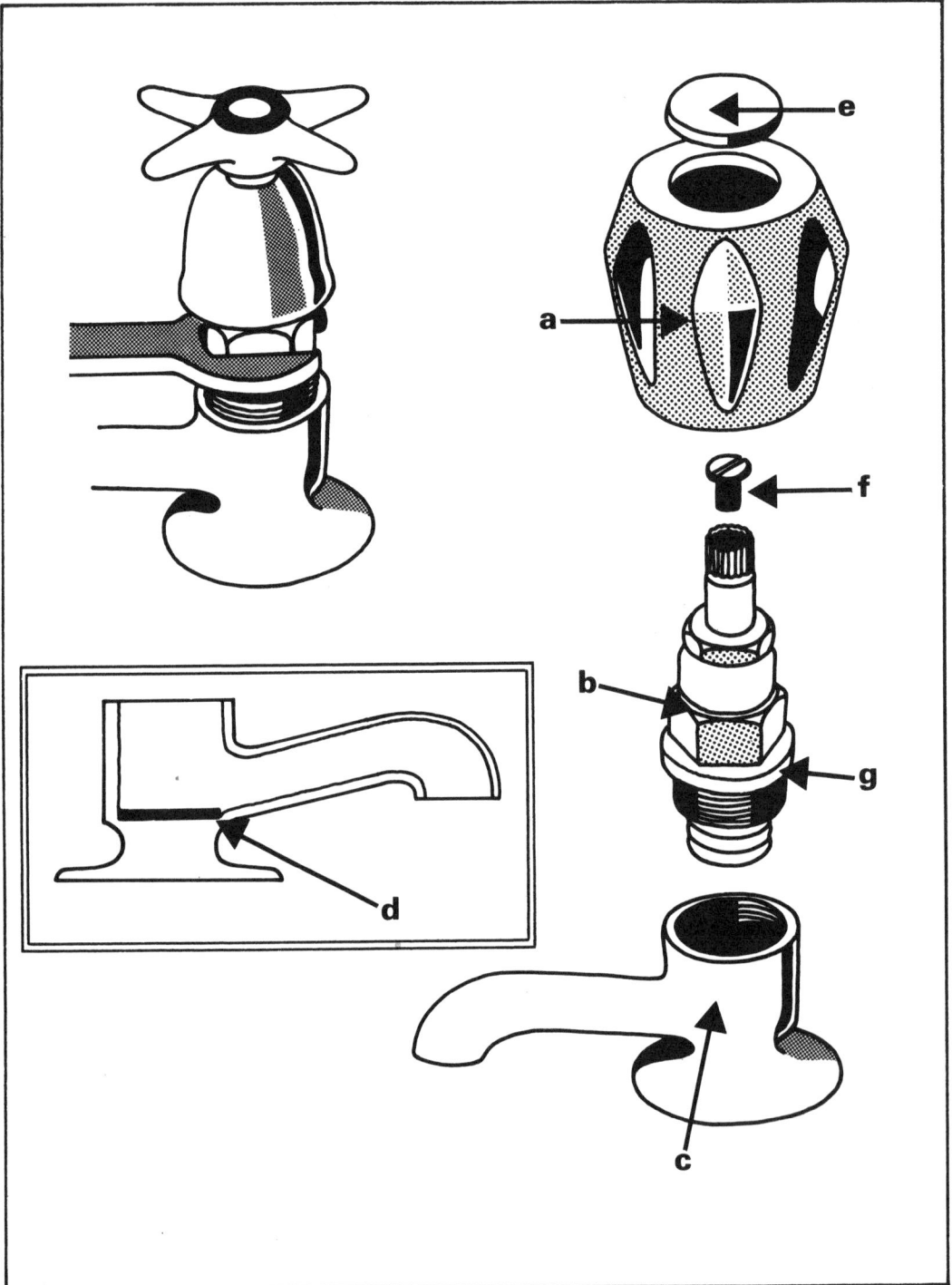

Fig. 4.

44

remaining cool even when the tap is discharging very hot water. For luxury water fittings, solid semi-precious onyx hand wheels are increasingly popular because of the beautiful grain displayed.

The 'Supatap' introduced by Deltaflow was and still is unquestionably unique. The design has found favour on innumerable occasions being an inspired piece of simplicity and restraint. Supatap maintenance can be carried out without shutting off the water supply because of its built-in automatic valve which cuts off by water pressure when the nozzle is removed. *(Fig. 5.)*

All taps and other fittings should be connected to conform with the regulations of the local water authority. The flow rate per tap as recommended in the BS Code of Practice (rate in litres per minute) is as follows:

Bath   — 18 l/min
Washbasin   — 9 l/min
Sink   — 11 l/min
Shower   — 7 l/min

Combination fittings should discharge almost double these volumes. In this context combination fittings are mixers into which hot and cold water are fed, being discharged from a common outlet. They are used for washbasins, baths and to a lesser extent for sinks. The fitting mostly used for sinks is a combination unit with dual flow, which is not a true mixer.

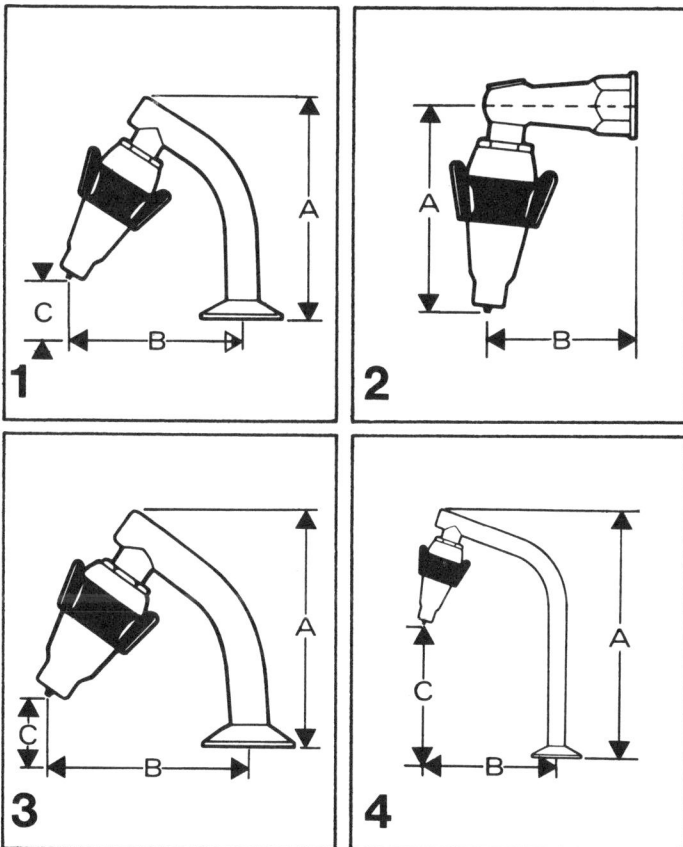

1 Basin tap
A: 108 mm
B: 92 mm
C: 29 mm

2 Wall tap
A: 84 mm
B: 67 mm

3 Bath tap
A: 118 mm
B: 102 mm
C: 27 mm

4 Sink tap
A: 197 mm
B: 118 mm
C: 106 mm

Fig. 5.

10 Traps, Waste Pipes and Overflow Pipes

Until comparatively recently it was common practice to take the pipes carrying waste water from toilets, baths, washbasins and kitchen sinks through the outer wall of the building to discharge into the underground drainage system. Waste from the toilet in such a system was flushed into a vertical soil pipe connected to the foul-water sewer often near to an inspection chamber (see Chapter 15). This vertical soil pipe was extended as a vent pipe beyond the point at which waste water entered, its highest point then being above any windows usually a little above the eaves. *(Fig. 1.)* Foul air can then escape without being objectionable. Other waste water from baths etc., was then discharged into a rainwater head in the down pipe from the gutters which in turn ran into the drains via a gully at ground level. External waste-water systems such as this are very vulnerable to frost in severe weather. The modern practice is the single-stack system where all waste water from sanitary appliances is discharged into just one internal pipe which in turn is connected to the underground drain. *(Fig. 2.)* In this way no waste pipes are exposed outside the house with the exception of that carrying rainwater. This too is sometimes run down the inside of walls particularly in newer public buildings.

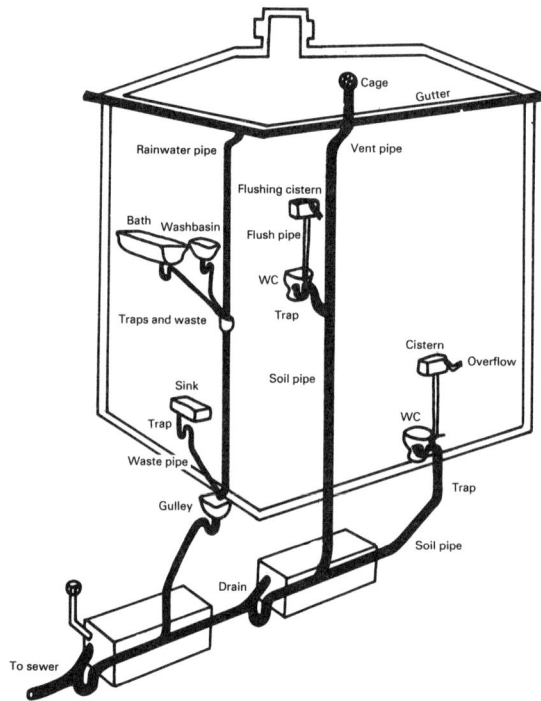

Fig. 1.

Any odours associated with waste water from toilets, bidets, baths, washbasins and sinks are prevented from passing back into the building by water-sealed traps. These may be P- or S-traps depending upon their shape, those for toilets being an integral part of the fitment, others being installed in the pipework. In older property these will invariably be of lead, but nowadays are more often of copper or plastic; the principle being that a small quantity of water will remain in the U-bend to form the seal. It should be emphasised that nothing which may cause the trap to become blocked should be discharged with the waste water from baths, washbasins, sinks etc. Any solid object accidentally allowed to do so will rapidly build up a blockage of small pieces of debris around it, which would otherwise have passed through the trap.

The different ways in which blockages may be removed without dismantling the trap are dealt with in Chapter 16, but should these be unsuccessful lead and copper traps have brass cleaning eyes set into them. These are at the bottom of the U-bend, usually at the sides of P-traps, and underneath S-traps. The plug(s) can be unscrewed with a wrench or may

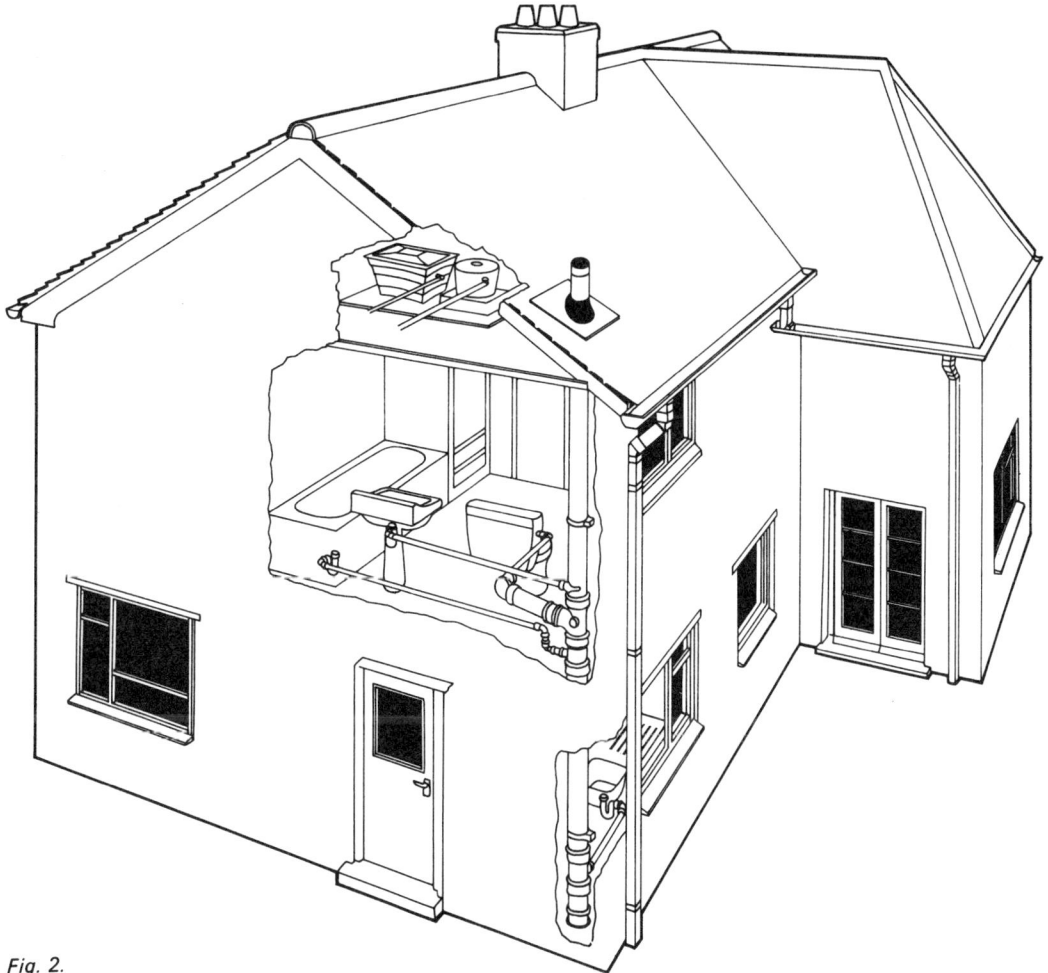

Fig. 2.

have slots in which a lever can be applied. Care must be taken to replace the hard rubber or fibre sealing ring and to give the pipework adequate support as the plug is removed or replaced.

Modern plastic pipework for waste water is specially formulated to withstand hot water and to resist the acids and detergents associated with domestic waste. *(Fig. 3.)* The traps designed for this system are swivel-necked to give maximum adjustment when being installed. The connection of the waste pipe is simplified because the direction of the outlet from the trap can be adjusted through 270 degrees to almost any desired position. In confined spaces such as those behind washbasins, pedestals or under sinks this is particularly useful. These plastic traps have excellent flow and self-cleaning properties, but when necessary the centre joint does provide easy access for the removal of blockages; BS 2494 part 2 specifies the high quality of the rubber seals required for plastic traps and pipework. The outlets of plastic traps are

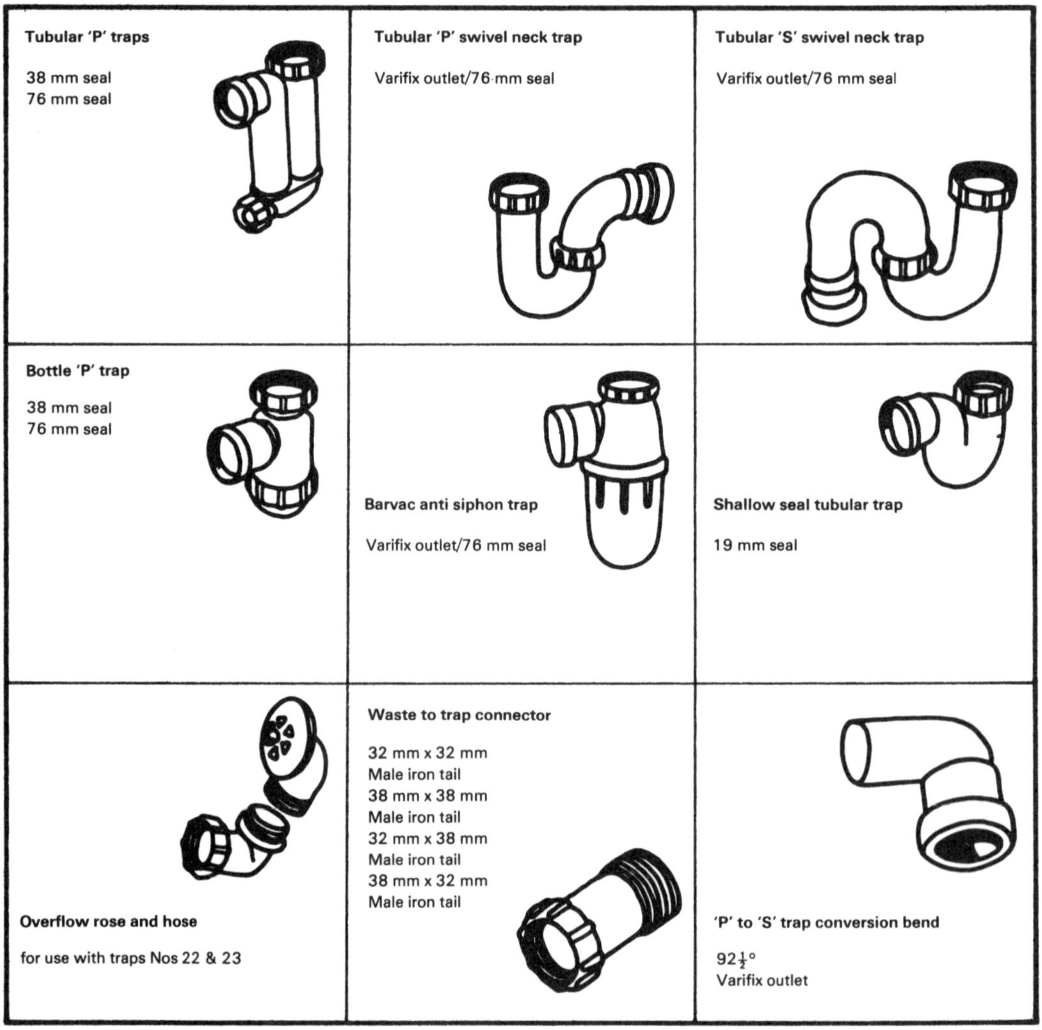

Tubular 'P' traps

38 mm seal
76 mm seal

Tubular 'P' swivel neck trap

Varifix outlet/76 mm seal

Tubular 'S' swivel neck trap

Varifix outlet/76 mm seal

Bottle 'P' trap

38 mm seal
76 mm seal

Barvac anti siphon trap

Varifix outlet/76 mm seal

Shallow seal tubular trap

19 mm seal

Overflow rose and hose

for use with traps Nos 22 & 23

Waste to trap connector

32 mm x 32 mm
Male iron tail
38 mm x 38 mm
Male iron tail
32 mm x 38 mm
Male iron tail
38 mm x 32 mm
Male iron tail

'P' to 'S' trap conversion bend

$92\frac{1}{2}°$
Varifix outlet

Fig. 3. Traps and fittings. 32 and 38 mm bore

Principles of operation

a. Shows the Barvac trap under normal operating conditions with full water seal.

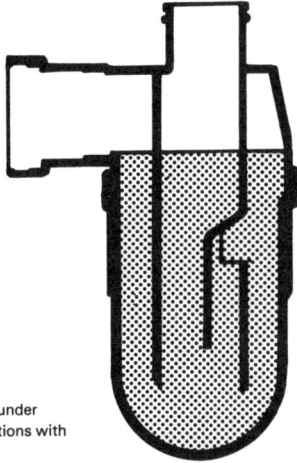

b. When subjected to severe siphonage conditions the Barvac automatic hydraulic action allows air through the by-pass tube without any major loss of water.

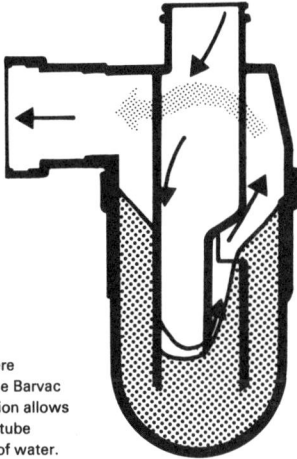

c. When normal conditions return the remaining water falls back to re-seal the trap – 85 mm deep, minimum 40 mm

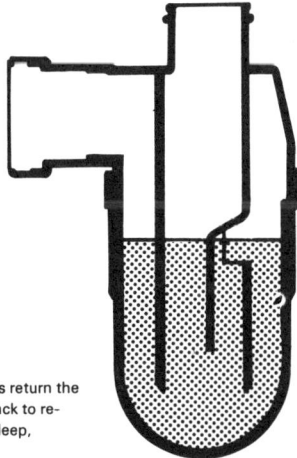

Fig. 4.

usually designed to accept imperial and metric copper tube, push-fit and solvent-welded plastic pipework.

The water seal in traps can be broken under certain circumstances, e.g. if a washbasin empties with such rapidity that all water is drawn away with it, or exceptionally strong winds suck out the water from the trap where the outlet is outside the building. The introduction of the single-stack system created further problems when it was found that in order to preserve the water seal any 32 mm installation, e.g. a washbasin, or 38 mm, e.g. baths and sinks, had to be no more than 1·675 m and 2·286 m respectively from the stack even though special deepseal traps were used. Failure to observe these distances resulted in loss of seal, but the problem has now been overcome by the development of the anti-siphon bottle trap. This allows the architect or plumber to position fitments wherever they are required within a building with no risk of siphoning off the water seal. *(Fig. 4.)*

The traps for WCs are an integral part of the fitment, the U-shape being formed in the pan and the water it holds forming the seal. The outlet then connects with the soil pipe. The majority of WCs are wash-down types, and when flushed water from the cistern is discharged down the flush pipe. This runs inside the rim of the pan and flows over the inner surface. When a sufficient head of water has been built up the contents of the pan are displaced, being flushed through the soil pipe forward to the drains. This leaves the pan clean and the trap recharged with water, the waste matter being cleared by the volume and velocity of the water. Other WCs work by siphonic action and the pans can be single trap or double trap. The outlet of the single trap is so formed to impede the flow until the bend above is full of water when the siphonic action begins. In the double-trap type an air pipe connects the space between the two traps. Air is sucked out through this pipe as the water rushes down the flush pipe assisting the siphonic action *(Fig. 5.)*

Extreme care must be taken to avoid allowing anything to fall into the pan of a WC which could cause a blockage. This is particularly important with the siphonic types.

To replace an outdated lavatory pan is not too difficult, remembering that as well as being of different kinds, they are available with different spigots, i.e. the drain end which connects to the soil pipe. Both washdown and siphonic pans may be

Wash down WC

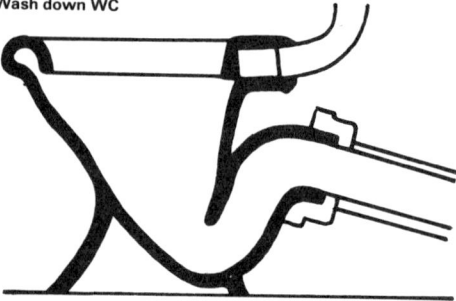

Siphonic WC, single trap type

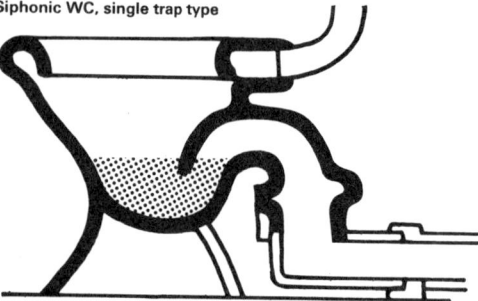

Siphonic WC double trap

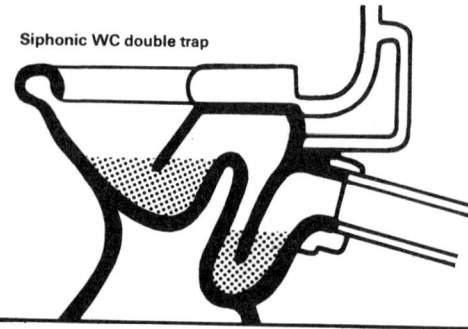

Fig. 5.

obtained with the spigot so inclined to connect with soil pipes passing through the wall or down through the floor. Check to get the right type, matching the measurements of the new as closely as you can with the old, paying particular attention to the fitting of the spigot to the soil pipe. The outlet of modern P-type pans is standard, the centre being 190 mm from the floor. If a low-level cistern is being fitted instead of a high one, the pan may need to be further from the wall, necessitating the use of extension pipes. Your plumber's merchant will advise on what is available. Failing this, slim cisterns are made which protrude only 100 mm or so from the wall. However, in order to contain the amount of water necessary to flush the toilet, these are of a larger area and take up more wall space.

Modern low-level cisterns are screwed to the wall and are so designed that the brackets which support them are hidden. It may also be necessary to make a hole through the wall for a new overflow pipe. The procedure to adopt in renewing a WC is as follows, with some possible minor amendments for individual cases:

1. Flush the pan clean, turn off water at stopcock and empty the cistern by flushing again. Mop out the trap with rags.
2. Take out the screws holding the pan to the floor; these should be of brass to avoid corrosion.
3. Disconnect the flush pipe – this may be a simple finned rubber seal, or a clamped rubber seal wrapped round a rubber sleeve.
4. The ideal is to release the pan so that it can be removed without damaging the soil pipe. It is at this stage where most care is required, much depends upon how the joint has been made. Most difficulty is likely to be experienced where the WC has been mounted on a concrete floor, the soil pipe cemented in vertically and the joint between the

two made with cement. In this case, to avoid damage to the socket of the soil pipe, the pan spigot may have to be shattered with a hammer. Binding first with a strong adhesive tape will prevent broken pieces falling down the soil pipe. Once the pan can be lifted clear, the soil pipe should be stuffed tightly with a rag to prevent other pieces falling in as the socket is cleared. The rag also blocks off unpleasant odours.

5. A small, sharp cold chisel and light hammer should be used to chip out remaining waste, working inwards and away from the soil pipe — damage to which must be avoided at all costs.

This joint will most likely have been made with paint and putty if the WC was mounted on a wooden floor, so that it is resilient to allow for movement of the timber (this could be due to expansion or contraction of the wood, or slight flexing of the boards as the floor is walked on). If the floor is solid concrete, then the joint may have been made with a mortar of one part cement to three parts of sand, but in each case the joint will first have been packed with hemp before being pointed off with putty or mortar. As an alternative to these methods the new pan may be joined to the soil pipe by means of a patent connector such as Bartol's 'Polyflex'. Basically these are flexible sleeves which allow for up to five degrees of misalignment and where necessary will mould themselves to misshapen spigots. The actual connection is a simple push-fit. Because of the different outside diameters of pan spigots, a range of connectors is made.

6. If the cistern is being replaced, this should now be screwed to the wall on its brackets. The height is important so that the flush pipe reaches down to its connection on the pan.

7. Try the new pan in position, check the fitting of the flush pipe and mark for the fixing screws. In a wooden floor, a pilot hole for each screw can be made with a bradawl, but a concrete floor must be drilled and plugged. A rotary percussion drill such as a Wolf 3548 (see *Home Maintenance and Outdoor Repairs* in this series) is ideal for this purpose, the holes being drilled to accept fibre or plastic plugs. Remember to use brass screws of a suitable length and avoid overtightening which may crack the pan. A pan which is slightly too low to meet the soil pipe can be lifted a little on a board cut to the profile of the base, but keep this to a minimum since the height of the pan is designed to suit persons of average build. Pans being set on concrete should be bedded on a mortar of one part cement to four parts sand, pointing it off neatly round the base.

8. Now connect the flush pipe — the fittings with which to do this will have been supplied by the manufacturers (see Chapter 12).

9. Make off the joint between spigot and soil pipe — first pack with hemp and ram down firmly, then fill with mortar or putty, depending upon requirements (also see *5.* above). This should be pressed well in to seal the joint and pointed off to a neat bevel. A putty joint should be allowed to stand several days and be painted to prevent it from crumbling.

10. Reconnect the cold supply and overflow pipes to the cistern using suitable washers and jointing compound. Turn on the supply once more at the stopcock, flush the pan and check for leaks.

Waste from WCs is discharged into the soil and vent system of pipework. These are now almost invariably of plastic with push-fit joints facilitating easy on-site

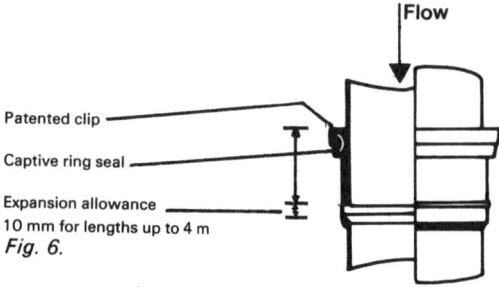

assembly. The joints are made efficient by a specially designed synthetic rubber sealing ring contained in a moulded housing which prevents displacement of the ring as the pipe end is inserted. *(Fig. 6.)* The spigot of the fitting or pipe to be joined must be clean, the end bevelled at 150 degrees and lubricated. Silicone lubricant in aerosols is available for this purpose. The end of the pipe must be fully inserted and then withdrawn 10 mm. This provision for expansion and contraction because of thermal changes is adequate for

Flow

Patented clip

Captive ring seal

Expansion allowance
10 mm for lengths up to 4 m

Fig. 6.

WC connectors and access fittings

Straight WC connectors

Bent WC connectors

"Polyflex"
WC connectors

Access bend

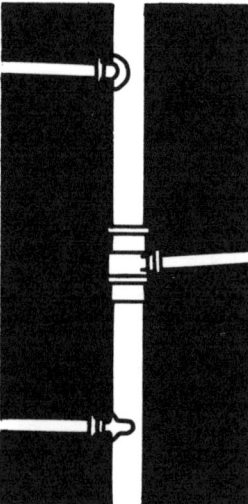

Boss fittings

Patch boss

Solvent boss adaptors

Grating, reducer and weather fittings

Balloon grating

Roof weathering slate

Fig. 7a.

lengths up to 4 m and must be maintained by fixing the pipe at its upper end with a socket bracket. Provision for expansion must also be made at the foot of the stack by using a drain connector, the top of which must always be above finished floor level so that there is free movement in the expansion socket. Connectors are available to adapt to clay drains or cast iron.

Where the vent pipe passes through the roof a multipitch weathering slate is used. These are in three slate sizes and are suitable for roof pitches from 10 degrees to 55 degrees. A rubber cone fits snugly round the pipe, and the malleable aluminium base can be dressed easily to the profile of the slates. A balloon grating prevents birds from getting into the pipe.

Other waste pipes can be run into the soil pipe by means of a range of solvent-welded bossed fittings which incorporate a similar push-fit principle. Suitable holes are drilled with a hole cutter and the boss fitted with gap filling solvent cement. *(Figs. 7 a and b.)*

These waste pipes are mainly from

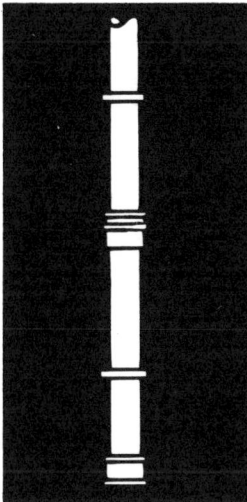

Straight pipe and sockets, drain connectors and fixing brackets

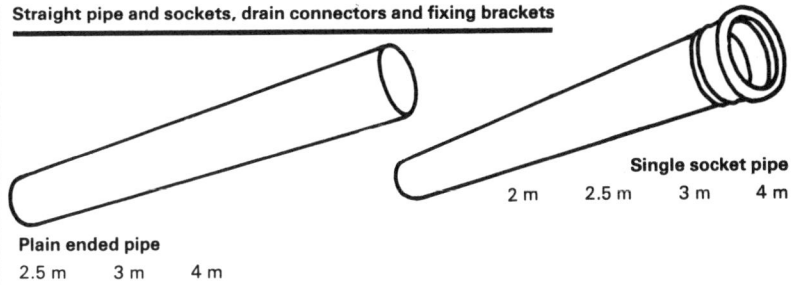

Plain ended pipe
2.5 m   3 m   4 m

Single socket pipe
2 m   2.5 m   3 m   4 m

Single socket

Drain connector
to Clay

Pipe fixing bracket
Quickfix

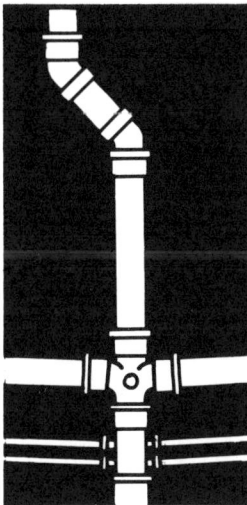

Bends and branches

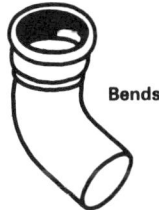

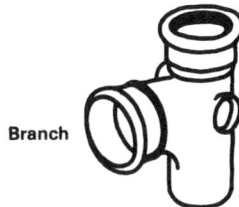

Bends

Double branch

Branch

Unequal branch

Fig. 7b.

53

washbasins and baths and are 32 mm, 38 mm and 50 mm bore with pipe-wall thicknesses of 1·8 mm, 1·9 mm and 2 mm respectively, each supplied in 3 m lengths. *(Fig. 8.)* They may be a push-fit system as are soil pipes, or solvent-welded (Chapter 6). The plastics from which they are made are not affected by boiling water and are resistant to those acids and detergents in domestic use. Installation is simple, only a small toothed saw is required to cut the pipe and a file or pocket knife to bevel the

Fig. 8. Push-fit waste systems. 32, 38 and 50 mm bore

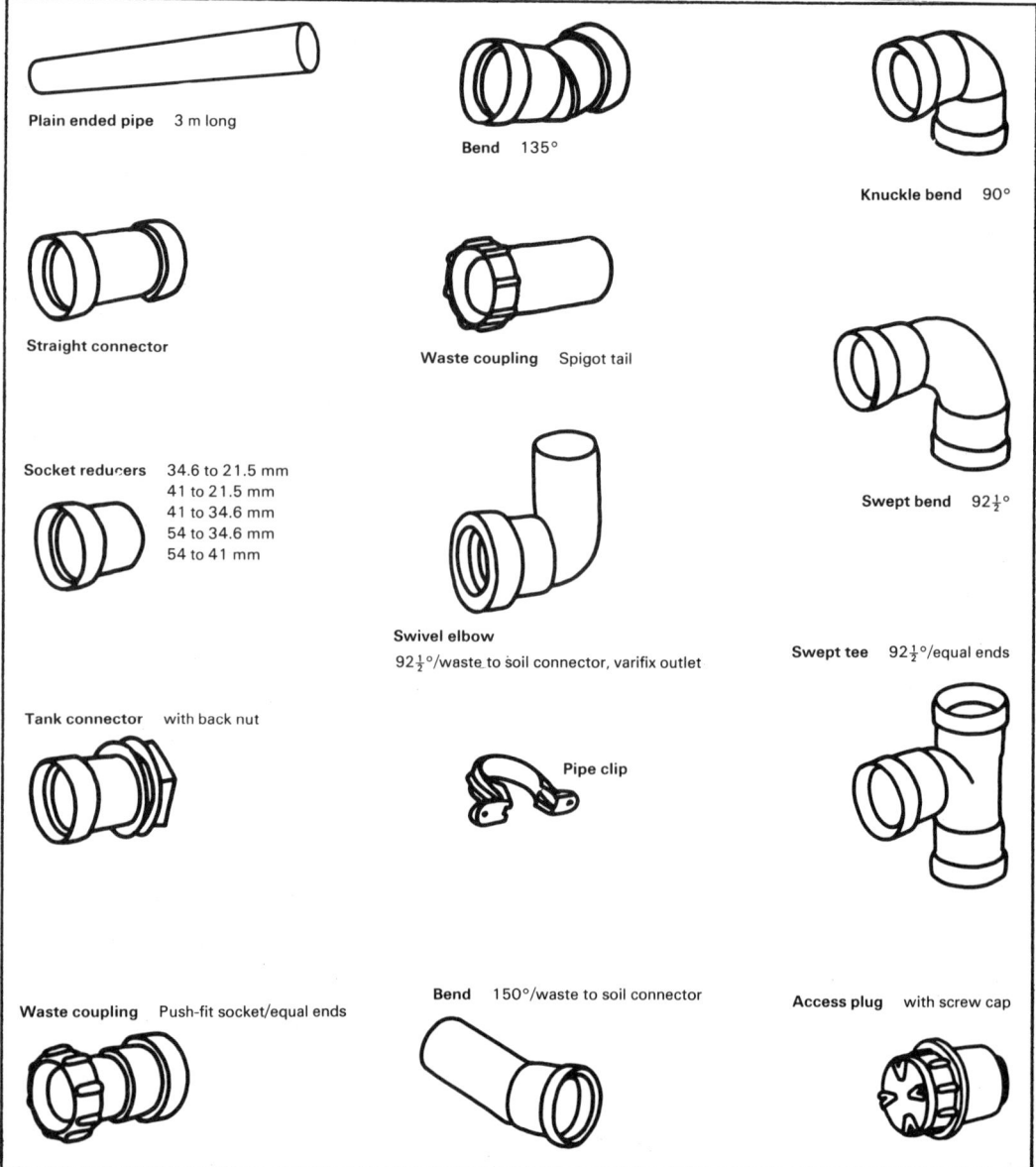

Plain ended pipe   3 m long

Bend   135°

Knuckle bend   90°

Straight connector

Waste coupling   Spigot tail

Swept bend $92\frac{1}{2}°$

Socket reducers   34.6 to 21.5 mm
41 to 21.5 mm
41 to 34.6 mm
54 to 34.6 mm
54 to 41 mm

Swivel elbow
$92\frac{1}{2}°$/waste to soil connector, varifix outlet

Swept tee $92\frac{1}{2}°$/equal ends

Tank connector   with back nut

Pipe clip

Waste coupling   Push-fit socket/equal ends

Bend   150°/waste to soil connector

Access plug   with screw cap

end.

Expansion is allowed for in the sockets of the push-fit system, whereas expansion connectors must be used at intervals in the solvent-welded system to provide for thermal movement. Where it is desirable to prefabricate waste pipes prior to installation the solvent cement method of jointing is particularly suitable. *(Fig. 9.)* The joining of traps to waste outlets is by compression nut and washer; the nut should not be overtightened.

Fig. 9. Solvent-weld waste systems. 32, 38 and 50 mm bore

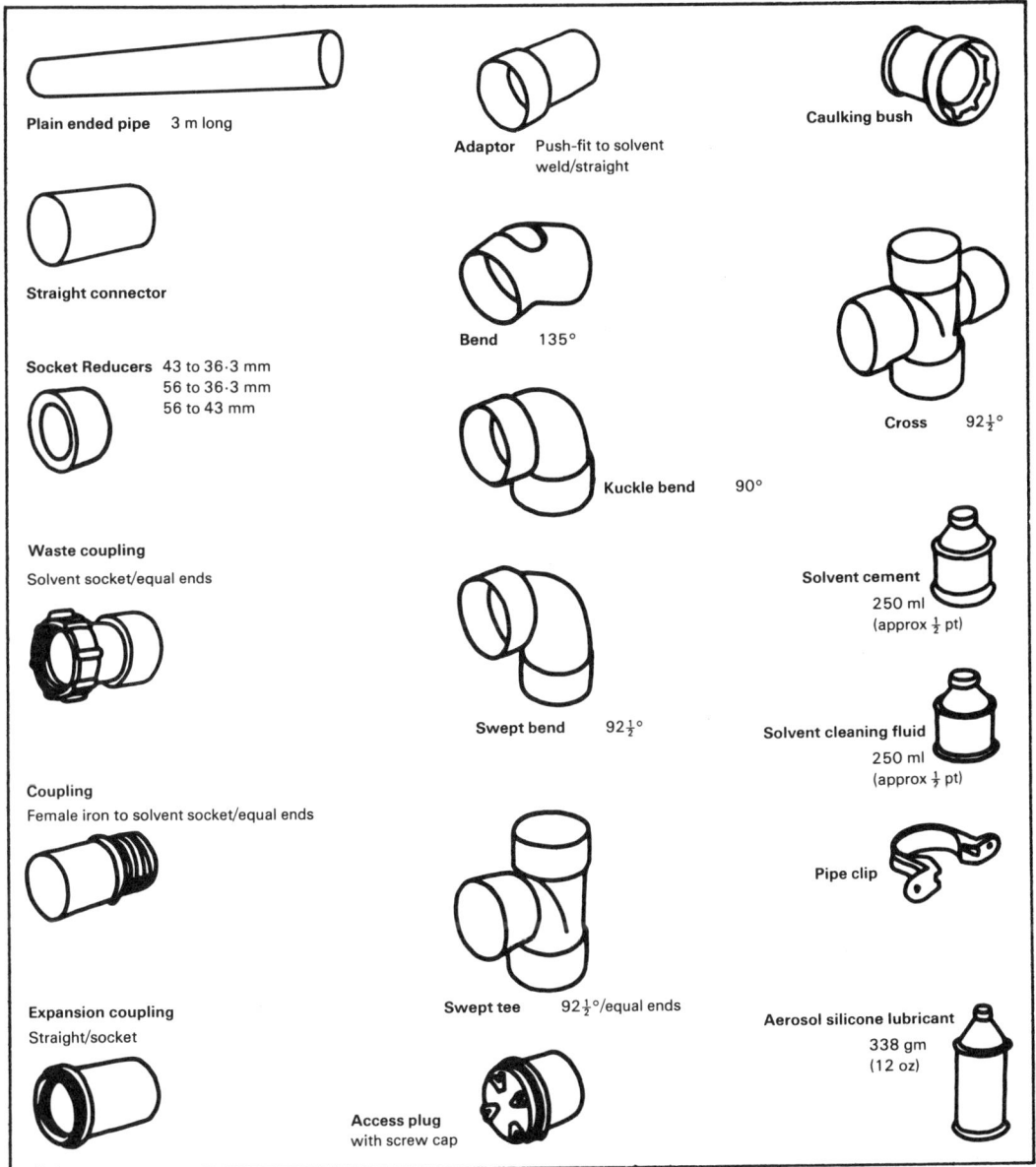

Plain ended pipe   3 m long

Straight connector

Socket Reducers   43 to 36·3 mm
56 to 36·3 mm
56 to 43 mm

Waste coupling
Solvent socket/equal ends

Coupling
Female iron to solvent socket/equal ends

Expansion coupling
Straight/socket

Adaptor   Push-fit to solvent weld/straight

Bend   135°

Kuckle bend   90°

Swept bend   92½°

Swept tee   92½°/equal ends

Access plug
with screw cap

Caulking bush

Cross   92½°

Solvent cement
250 ml
(approx ½ pt)

Solvent cleaning fluid
250 ml
(approx ½ pt)

Pipe clip

Aerosol silicone lubricant
338 gm
(12 oz)

Cisterns fitted with ball valves must have overflow pipes connected which discharge outside the building. Push-fit systems·with 19 mm bore are ideal for this purpose, the connections to cisterns and terminal fittings being with compression nuts. *(Fig. 10.)*

Support brackets	Maximum distance between centres		
	Bore	Horizontal	Vertical
Soil and Vent Waste, Push-fit and Solvent-welded	75, 100 mm 32, 38, 50 mm	1·00 m 0·76 m	2·00 m 1·22 m
Overflow	19 mm	0·76 m	1·22 m

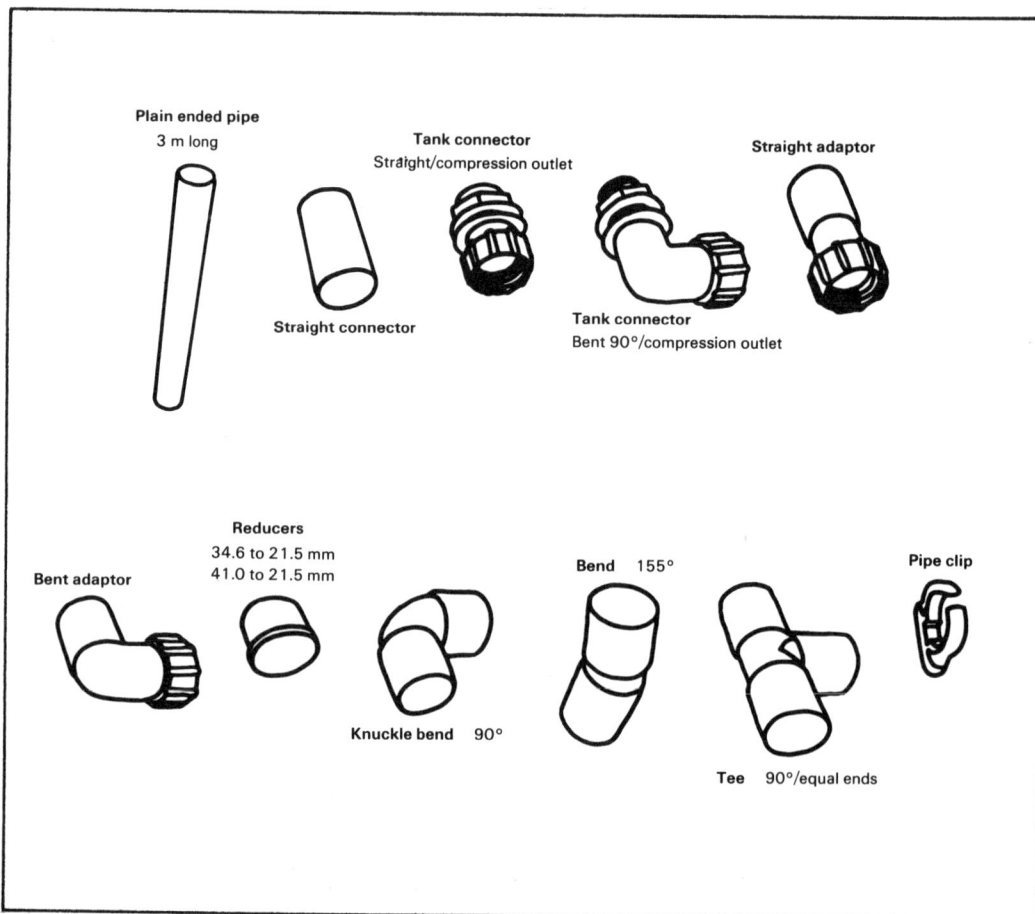

Fig. 10. Push-fit overflow system

11 Modern Cisterns

Cisterns are those plumbing fitments designed for the storage of cold water. Within the domestic system they are confined to the cold-water supply to baths and washbasins, expansion tanks for central heating systems and the means by which WCs are flushed. Over the centuries cisterns have been constructed from many materials, including stone and slate. For many years those for bathroom and central heating systems were of galvanised steel. It is now usual to install modern plastic ones which have the advantages of being free from corrosion, light in weight and robust in use. Good quality polypropylene cisterns as those manufactured by Bartol Ltd., are manufactured to BS 4123 and approved by the National Water Council. A variety of shapes and sizes are made *(Fig. 1.),* with capacities from 18 litres to 227 litres. These are injection moulded from polypropylene and as they are unaffected by discharges of hot water may also be used as expansion tanks. A further range of circular cisterns, with capacities of 114 litres and 227 litres, are made for cold-water storage only. *(Fig. 2.)* Matching lids are available for all of these to form a safe dustproof seal.

Recommended fixing instructions are as follows:
1. Ensure that the cistern is received free from damage.
2. Drilling
a. A hole saw should be used to produce a perfectly circular hole.
b. The cistern wall must be supported while drilling, i.e. with a piece of timber held behind it.
c. Heated metal tubing is not recommended.

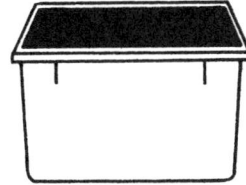

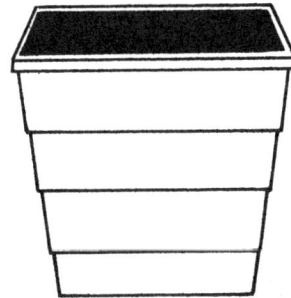

Fig. 1.   Cold water storage and expansion cisterns

Fig. 2.   Circular cold water storage cistern

3. Positioning.

d. A level, firm support must be provided for the whole of the base.

e. It should not be positioned close to electric light bulbs or any source of concentrated heat.

4. Assembling

f. Pipework must enter at 90 degrees to the cistern wall being so aligned to avoid distortion.

g. All pipes must be firmly fixed and supported independently of the cistern.

h. Do not overtighten backnuts.

i. Do not use jointing compounds. For sealing the connector use PTFE tape and polyethylene washers between the cistern wall and the flanges of the tank connectors.

j. The ball-valve plate, which is supplied loose, must be fitted to the outside of the cistern with the lock nut. The centre for this hole should be 38 mm from the rim of the cistern.

k. If the cistern has to be distorted in order to pass it into the loft do not drill until this has been done. The hole would create a weak point at which splits may occur. For the same reason care must be taken to avoid deep scratches.

Cisterns for toilets may be of vitreous china or plastic and may be low-level or close-coupled. (Fig. 3.) They are designed to hold 9·1 litres of water and may be of conventional dimensions or 'Fineline'. These are used where limited space requires the cistern to project the minimum amount from the wall, usually about 110 mm. However, because of the necessary capacity they are therefore larger in area. WC cisterns must conform to BS 1125 (1973) and be suitable for use with WCs to BS 1213 (1945). They should have easily cleaned surfaces and a minimum of joints or mouldings where dirt can collect; they may be fixed by screwing to the wall, or supported on concealed brackets, or by a combination of both.

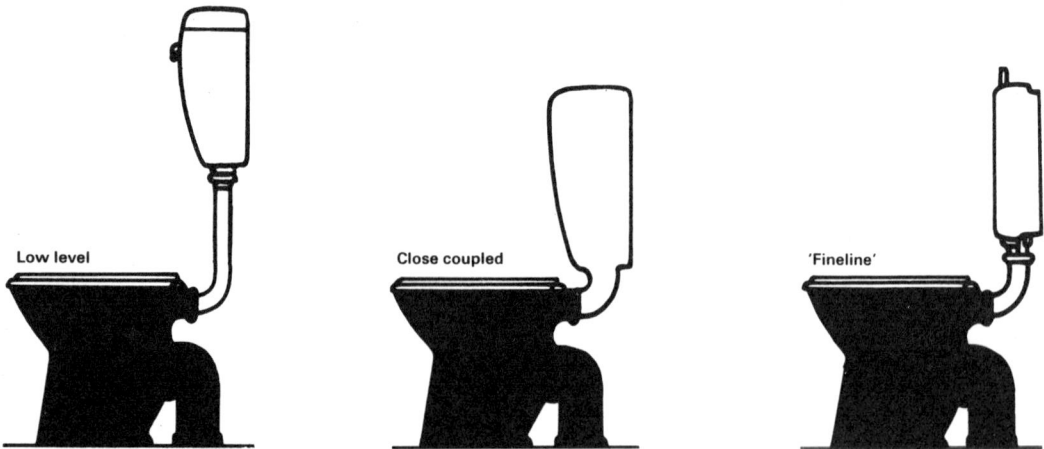

Low level   Close coupled   'Fineline'

Fig. 3.

12 The Modern Bathroom Suite

As it became customary for each house to have a bathroom, the fitments were often very austere and limited in number, often only a bath and washbasin with a separate WC. Depending upon the available space it is now not uncommon to have a bath, one or perhaps two washbasins, a WC and more recently a bidet. Of late, the shower cubicle (Chapter 13) has become a further addition as its value becomes increasingly appreciated, although it may not necessarily be sited in the bathroom. All too often the bathroom tended to be a somewhat cold, bleak room with a rather clinical atmosphere, and with little in the way of comfort. Modern materials and approach to design have brought about a revolutionary change, now making it possible for the bathroom to be one of the most stylish, pleasant and enjoyable rooms in the home. Whether you are planning a new bathroom, or giving an old one a facelift, you should consult the many manufacturer's brochures which are readily available. There is a quite bewildering array of styles and colours requiring careful consideration so that finally the right choice is made. For guidance, Table 1 gives an indication of the approximate amount of space occupied by a fitment and the necessary amount of space around it to permit easy movement. However, since it is unlikely that every fitment will be in use at the same time, these access areas may overlap to some extent. It is assumed that each fitment stands against a wall, and the distance it protrudes is then its 'projection'. All dimensions are in mm.

With these measurements as a guide, if you cut out rectangles of paper in proportion they can be fitted onto a plan of the room. This must, of course, be drawn to the same scale and the pieces, which should be labelled, or of different colours, moved around on it until the most suitable positions are found. You must, naturally, take into account the position of existing features such as doors, windows, cylinder cupboard, radiators, etc.

Probably the most significant factor in bathroom modernisation is the successful use of acrylic plastic from which to make baths, outdating the traditional cast iron or pressed steel. Both were cold, vulnerable to rust and the enamel chipped all too easily.

The advantages of acrylic fitments are numerous:
- Warm to the touch.
- Light in weight.
- Meet hot and cold service conditions favourably.

Table 1

Fitment	Average length against wall	Projection	Access area length width
Bath	1500–1800	690–790	1200 X 750
Washbasin	650	500	650 X 600
WC	450–500	700	500 X 600
Bidet	360–400	600	600 X 600
Shower	775	770	775 X 700

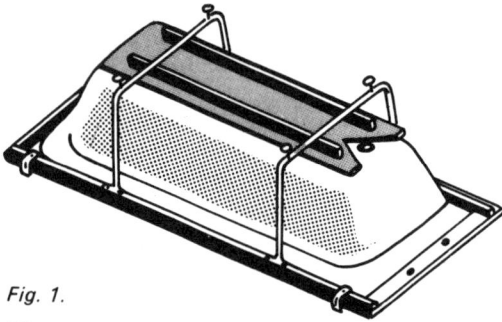

Fig. 1.

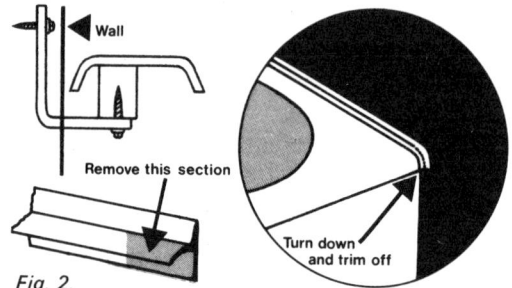

Fig. 2.

■ Impact strength conforms to BS 4305 (1972).
■ Resistant to most common household chemicals (paint-stripper, nail varnish and some dry-cleaning agents are exceptions).
■ Surface has natural anti-slip properties when wet.
■ Tough, resilient and durable.
■ Easily cleaned by rinsing with soapy water immediately after use.

The material is readily moulded enabling many alternative designs to be manufactured, all of which ensure comfort in use. Baths are formed from acrylic sheet 8 mm thick, or, depending upon their design, from 3 mm sheet with a bonded fibreglass (GRP) backing for strength. Accidental scratches can be removed with a fine abrasive, e.g. wire wool, followed by repolishing with metal polish. Scorching from cigarettes can also be remedied in the same way. This can be done without loss of colour.

Being a resilient material acrylic baths are delivered bonded to a cradle or with a supporting framework to which metal legs and feet are attached. *Fig. 1.* illustrates the method employed for the Chloride-Shires Symphony (see cover picture) and *Fig. 2.* the means by which it is anchored to the wall.

Installation is as follows:
1. Push tubular legs in to sockets on edge battens.
2. Fix legs in to notches on baseboard with screws.
3. Screw on wall brackets at required positions.

4. Fit taps and waste pipes; use resilient washers between nuts and bath, with sealing compound under waste flange; apply only moderate pressure when tightening nuts.
5. Place in position and level at a height of 500 mm using adjusting feet; lock the feet and screw to floor.
6. Screw fixing brackets to wall.
7. Seal between bath and wall with sealing strip supplied; fix with contact adhesive. *(Fig. 2.)*

Washbasins, WCs and bidets are generally of matching vitreous china because of the need for rigidity, and the use of household cleaners for hygiene, although some washbasins are of acrylic when used in vanity units. Many washbasins are screw-fixed to the wall and supported on a pedestal. They can be fixed on wall brackets in place of the pedestal if this is desirable, but the advantage of the pedestal is to conceal the pipework. Washbasins are usually moulded with anti-splash front rims, soap recesses and slotted overflows at the rear. Holes are provided for standard $\frac{1}{2}$ in taps or in some models for mixer taps.

WCs may be washdown or syphonic, low-level or close-coupled (Chapter 11). The pans are screw-fixed to the floor, close-coupled cisterns being supported by and bolted to the pan.

Bidets too are screw-fixed to the floor, with 'rim-supply' and 'pop-up' waste plug, and are ideal for full personal hygiene.

Vitreous china fitments should conform to BS 3402 (1969), the surface being resistant to most household chemicals and cleaning agents.

13 The Shower Unit

Whereas a short time ago a shower may have been considered as a luxury item in addition to the bath, people have been quick to appreciate its advantages. It is hygienic; the installation is comparatively simple and it is economical in use. A shower may be had using approximately one-sixth of the water needed for a bath, conserving both water and energy. A minimum flow rate of 3 to a maximum of 9 l/min are considered adequate; 4 to 7 l/min being a good working average. The efficiency of these flow rates is dependent on the discharge rose being well designed. A bewildering array of types of shower are available, ranging from a simple moulded rubber hose which connects to the bath taps — the correct mix being obtained by adjusting the hot and cold taps, to the modern, sophisticated thermostatic shower. This has two controls, one to select the flow-rate and the other the required temperature. It is the old story of paying your money and getting what you pay for. More satisfactory than the moulded rubber attachment are the 'panel bath mixer', the 'pillar bath mixer' and 'deck bath mixer' with diverters for shower fitments. *(Fig. 1.)* To taps such as these shower units similar to those in *Fig. 2.* may be attached. These are exposed units as opposed to the shower arm *(Fig. 3.)*, the supply pipe to which is concealed in the wall. The efficiency of cistern-fed shower units such as these depends upon the bottom of the cold-water cistern being not less than 1 m above the shower rose. *(Fig. 4.)* In order to achieve this it may be necessary to raise the level of the cistern in the loft on extra battens or trestles. If this is not possible then it may be necessary to install a pump into the system between the mixer and shower head.

With showers that are virtually an integral part of the bath, the water has to be contained, so that it is discharged down the bath-water system. Any walls against which the bath is set must be tiled, using a waterproofed grout. Open sides to the bath must be fitted with a plastic curtain or panels to direct the water down into the bath.

A number of kits are available, suitable for installation over a bath or in a separate shower cubicle. The latter immediately doubles the bathing facilities of the house, being particularly advantageous in that it need not be installed in the bathroom, but simply where there is sufficient space available and that all necessary connections may be made to it.

Water of an even, comfortable temperature can be obtained by the 'blended valve' principle, precise quantities of hot and cold water being drawn into the mixing chamber from where the blend of water required is dispensed. The valve, which may be concealed within the wall *(Fig. 5.)* or be surface mounted *(Fig. 6.)* may be fitted with shower arrangements similar to those in *Figs 2.* and *3.* This type of shower mixer is, however, non-thermostatic and is therefore only suitable for use where pressures are nominally equal.

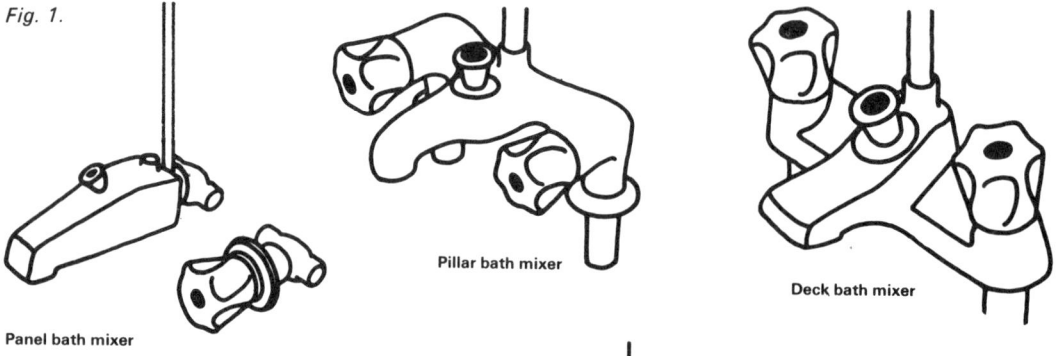

Fig. 1.

Pillar bath mixer

Deck bath mixer

Panel bath mixer

Fig. 2.

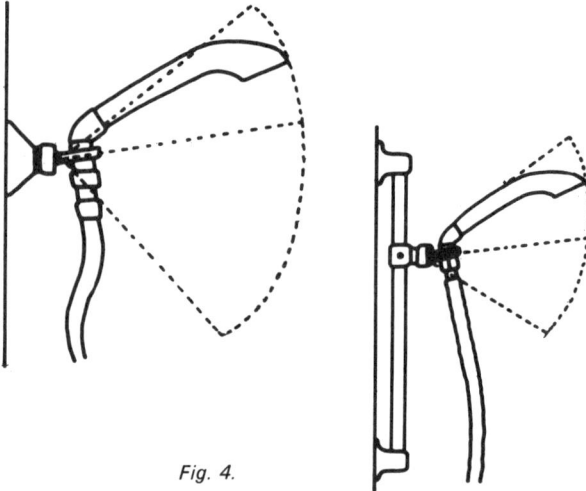

Fig. 3.

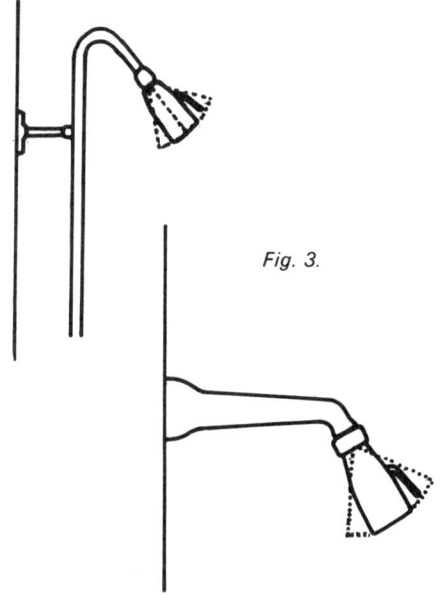

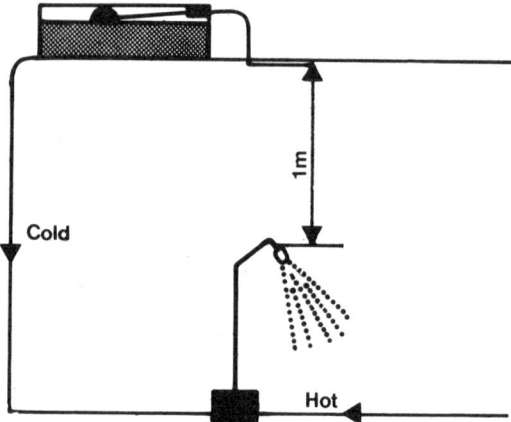

Fig. 4.

1m

Cold

Hot

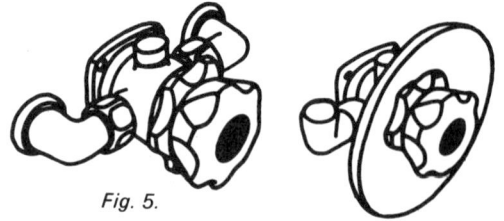

Fig. 5.

Fig. 6.

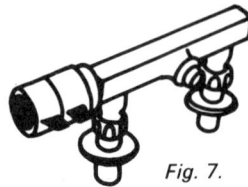

Fig. 7.

The addition of a thermostat is a further refinement *(Fig. 7.)*; the unit has two controls, one selecting the flow of water, the other the required temperature. The temperature is monitored and for safety, should the cold-water supply fail, an anti-scald device operates.

Whereas the previous kits rely on the plumbing system for their hot-water supply, packs are available which operate independently. Being so they may be sited at almost any desired part of the house. They operate from the cold-water supply, the unit's thermal element heating only the water which passes through as you take a shower *(Fig. 8.)* The saving of hot water is immediately obvious.

Complete kits are now available to enable you to build shower cubicles such as the Shires Osprey *(Fig. 9.)* fitted with an instantaneous water heater as above. This eliminates much extensive replumbing, thus reducing installation costs, as the only connections necessary are those to the cold supply and the electrical circuit. A waste outlet, usually in plastic, must also be run to dispose of the waste water. Cubicles require about 1 sq m of floor space, and once the position has been chosen, pipes for supply and waste should be run, finally moving the cubicle into place to make the final connections, the cubicle tray usually being supported on steel-framed cradles screwed to the floor. Access panels are provided to facilitate easy connections to the pipework and the heater has, of course, to be connected to the electrical supply. These same access panels also provide for easy maintenance (see also Chapter 10, and other chapters which cover the fitting of supply and waste water pipes).

Osprey shower cubicle

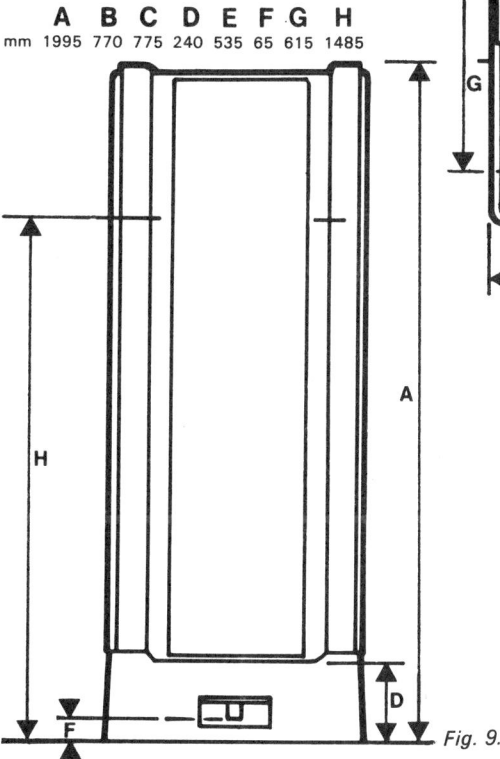

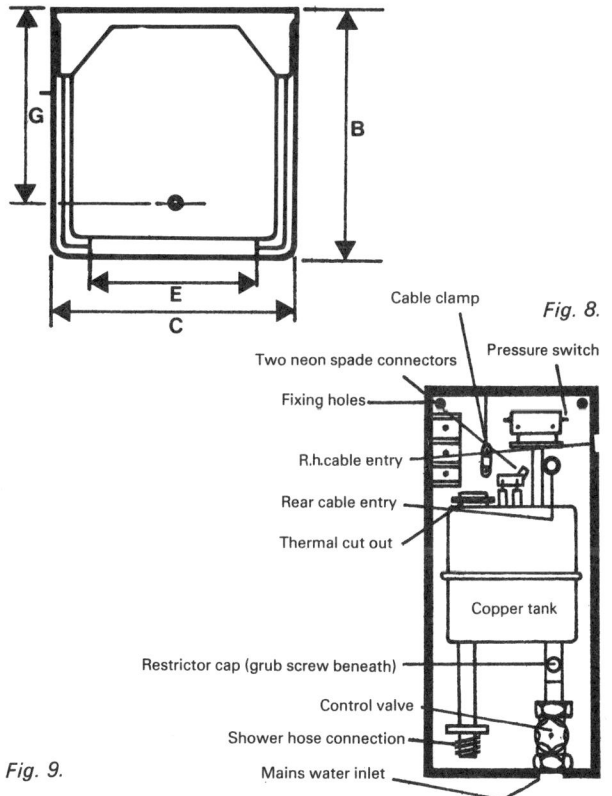

A	B	C	D	E	F	G	H
mm	1995	770	775	240	535	65	615	1485

Fig. 8.

Fig. 9.

Cable clamp
Pressure switch
Two neon spade connectors
Fixing holes
R.h.cable entry
Rear cable entry
Thermal cut out
Copper tank
Restrictor cap (grub screw beneath)
Control valve
Shower hose connection
Mains water inlet

14 Insulation of the Plumbing System and Frost Precautions

Conservation of energy in any form is acknowledged to be of world-wide importance. Conservation of heat in the domestic situation is economic commonsense, and assists in overall conservation. Insulation of the house is dealt with in more detail in other books of this series, whereas here we are basically concerned with plumbing. Suffice to say that lofts should be insulated with granular vermiculite or fibreglass between the joists. Both hot and cold-water pipes in basements, and ventilation areas under floors or unheated areas such as roof spaces should be lagged. So too should the hot-water cylinder; several patterns of cylinder jackets are available. Failing this the author has found that two or three layers of corrugated card fitted carefully round the cylinder and tied with string are very effective. Efficient insulation of the system not only conserves heat, but also guards against frost which can cause great inconvenience and expense. The volume of water increases by about one-tenth when it freezes, exerting great pressure, and this causes damage to, or bursting of pipes. Freezing obviously begins at the coldest point and depending upon its severity gradually extends along the pipe. Service pipes should not therefore be fixed to the inside of external walls, but be brought into the building at sufficient depth below ground until they can be brought up on an internal wall. Freezing will stop the flow of water and cut off the supply, but also if the pipes burst the escaping water may cause considerable damage to the structure of the building and its decor as the ice melts.

The cold-water cistern in the roof space must be lagged and this can be done by placing a loosely fitting box around it and filling the cavity with vermiculite or fitting slabs of expanded polystyrene round it. They can be held in place with string or a suitable industrial self-adhesive tape. As an alternative, fibreglass bonded mat may be used. A slab of polystyrene placed over the cistern will insulate the top and keep out dirt, but if any expansion pipes discharge into it provision must be made for this by leaving a suitable gap.

Some danger points for pipes have already been mentioned, but in addition pipes in outbuildings or garages, near air-bricks and ventilators or where cold draughts may occur should be lagged. Depending upon their situation this may be with pre-formed plastic foam, or pre-formed fibreglass both of which are normally supplied split to clip around the pipe. (Fig. 1.) Alternatively, thermal insulating material can be obtained in rolls to be wrapped around the pipe as a bandage. Different types are made of hair felt, polypropylene-backed jute mixture and fibreglass. (Fig. 2.). The recommended thickness for thermal insulation is 25 – 65 mm for pipes depending upon their diameter and situation, and 13 to 38 mm for cisterns. Insulation will delay heat loss and retard freezing, but if the temperature remains below freezing, further precautions may be necessary, such as placing a small electric or paraffin heater at vulnerable points. Electric heaters with porcelain enclosed elements, or 'black-heat' elements are best. If a paraffin heater is used it should be of a safety type with gauze round the flame.

Taps and overflows should not be

allowed to drip or waste pipes may freeze up. Because of this it is advisable to leave plugs in washbasins and baths during severe frost.

If any pipe is found to be frozen, place a small heater or even the electric iron near it, or cover it with a thick pad of cloth soaked in hot water on which further hot water can be poured from time to time until the pipe is thawed. Should the pipe be burst temporary repairs can be carried out by gently tapping the pipe back to shape then binding with plastic self-adhesive tape held in place with closely-wound strong string.

No attempt should ever be made to thaw pipes with a blowlamp, nor if the hot-water system is frozen should the boiler be fired, or in the case of a back boiler the fire lit in the hearth. A severe explosion could be the result in either case.

If the house is to be unoccupied for more than twenty-four hours in frosty weather and cannot be heated, the whole of the water system should be drained off. This means closing the service stopcock and opening all the draw-off taps, including drain taps for heating systems. Also, salt or anti-freeze solution should be added to the traps of WCs, baths and washbasins.

To refill the system, drain taps must be closed, the stopcock opened and finally all the taps closed once the water is flowing freely and all air has been driven out.

Preformed pipe insulation
For hot and cold water pipes. Self-extinguishing, lightweight and flexible. Operating temperature maximum 105°C (220°F), minimum minus 40°C (minus 40°F).
Supplied split for easy fixing but can be supplied unsplit in 2 m nominal lengths. Wall thickness 9 mm ($\frac{3}{8}$ in).
For 15 mm copper tube
For 22 mm copper tube or $\frac{1}{2}$ in ferrous pipe
For 28 mm copper tube
For 35 mm copper tube or 1 in ferrous pipe
For 42 mm copper tube
For 54 mm copper tube
  Tape 2 in x $\frac{1}{8}$ in In 100 ft rolls
  Adhesive 500 ml tin.

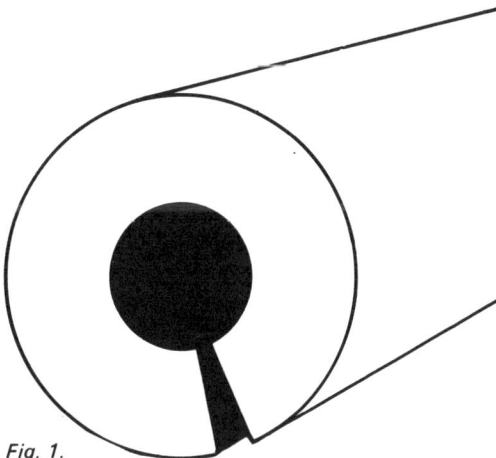

Fig. 2. Pipe insulation

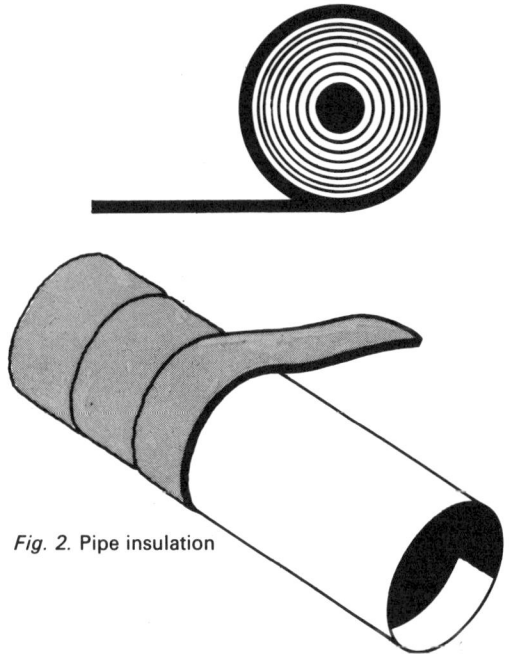

Fig. 1.

15 Drainage Information

Domestic drainage systems consist of above-ground work from gutters, sinks, baths and toilets connected to the below-ground system of drains. It is essential to know just how this is done for any particular house before considering maintenance or repair work. Whenever drainage is discharged into the system, foul air is prevented from escaping into the house by means of traps or water seals, the most common being the U-bend. *Fig. 1.* indicates the above-ground arrangement of external cast iron pipework typical of older property.

The so-called soil pipe carrying waste from the WC is extended above the eaves, or the level of any dormer windows as a vent for foul air. Surface water from gutters may also discharge into the foul-water sewer depending upon the requirements of the local authorities. If excess water of this nature might result in the local sewage works being overloaded, it may have its own sewer running into a stream or river. Should this not be convenient it may run into a soakaway situated quite near to the house providing the subsoil is sufficiently absorbent. Above-ground pipework for

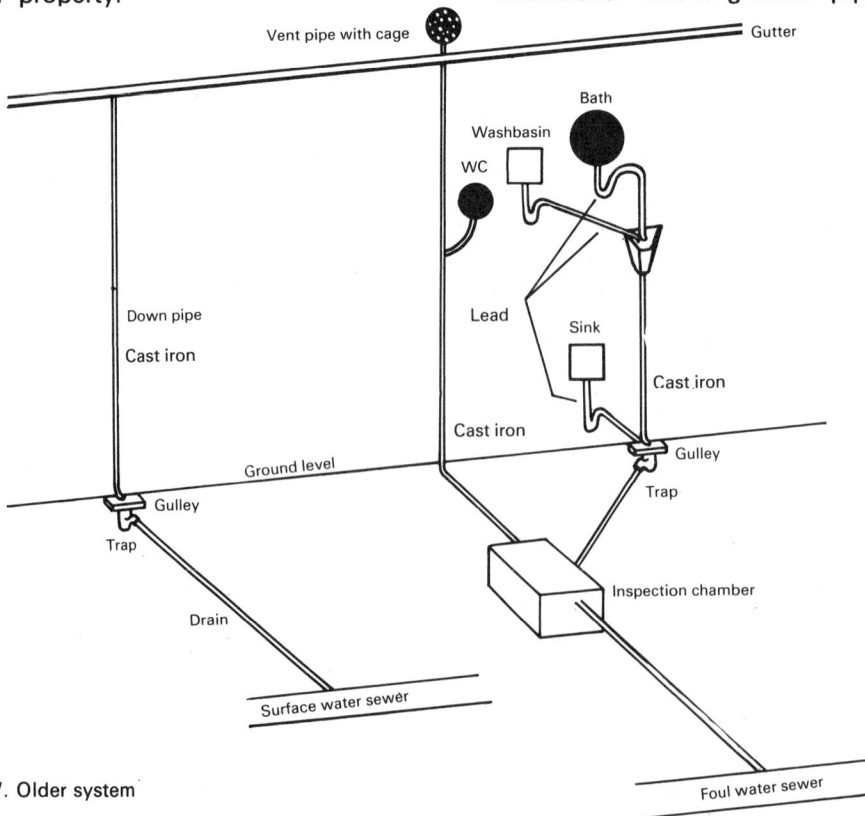

Fig. 1. Older system

66

such a system is invariably attached to the outside of the building; it is liable to freeze, requires costly maintenance and is aesthetically undesirable. This has led to the now compulsory modern system where a common soil and waste-water pipe (see *Fig. 2.*, Chapter 10) is constructed inside the walls of the house. In both cases, of course, rainwater is carried to the drains or soakaway by its own system of pipework. The soakaway should be fitted with a lid or concrete cover and consists of .a pit with porous walls, the cavity being open or filled with rubble.

Underground drainage pipes must 'fall' away from the building, so that water and waste matter run away freely. There will be manholes within the system at convenient points. These are to allow for inspection and for 'rodding' should blockages occur.

Gullies have a U-bend which forms a permanent water seal and are frequently used where waste-water pipes join the main drainage system. Modern gullies are designed so that the waste pipes enter either through the grid or below it by a side or back entry. In this way waste matter does not lodge on top of the grid, a cause of frequent overflowing. *(Fig. 2.)*

Underground drainage pipes were traditionally of glazed earthenware or cast iron, more recently of pitchfibre, and are now often of plastic. Where rainwater pipes connect to a combined drain a gulley must

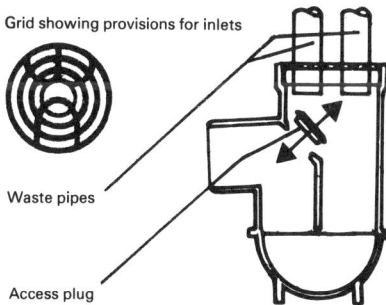

Fig. 3.

be used. The Plastidrain access gulley is ideal for this purpose since the connection is roddable from the gulley and avoids the need for an inspection chamber.

Where separate stormwater drains are provided a Plastidrain access gulley may still be used for the rainwater pipe connection in order to filter leaves from roof drainage in wooded districts or deal with surface water run-off from paved areas. *(Fig. 3.)* Otherwise, for connections to separate storm-water drains Plastidrain adaptors provide push-fit connections. The hole in the top face of the adaptor is offset to allow for the variations in distance between the down pipe and the building. *(Fig. 4.)*

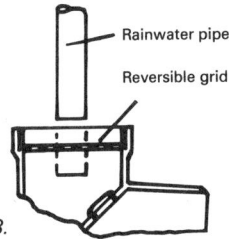

Fig. 4.

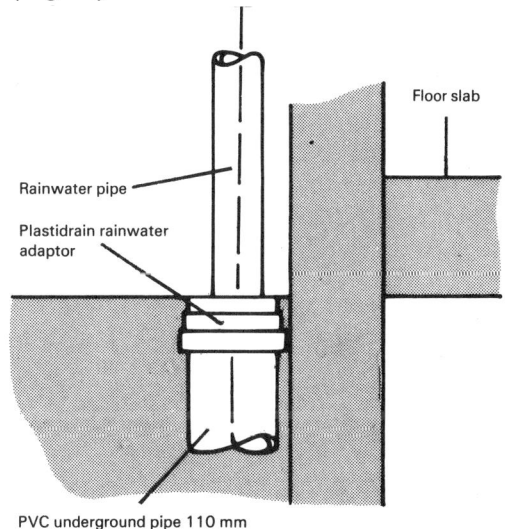

Fig. 2.

The inspection chamber is probably the most expensive part of traditional house drainage since it is normal to find all drains from the house connected to the sewer or collector drain via manholes which are generally 1·5 − 2·5 m deep. The high costs of excavating and constructing manholes, as well as the cost of site concrete in deep excavations near to foundations, can be avoided by the use of shallow Plastidrain glass reinforced plastic inspection chambers positioned off the main run of drain. *(Fig. 5.)*

An inspection chamber should be installed in the closest location permitted by the pipework connections and consistent with rodding accessibility. Placed this way, the chamber will usually need to be 610 mm deep, as will all the drains from the house. *(Fig. 6.)* In more complex designs the last chamber before the collection drain should be no deeper than 910 mm.

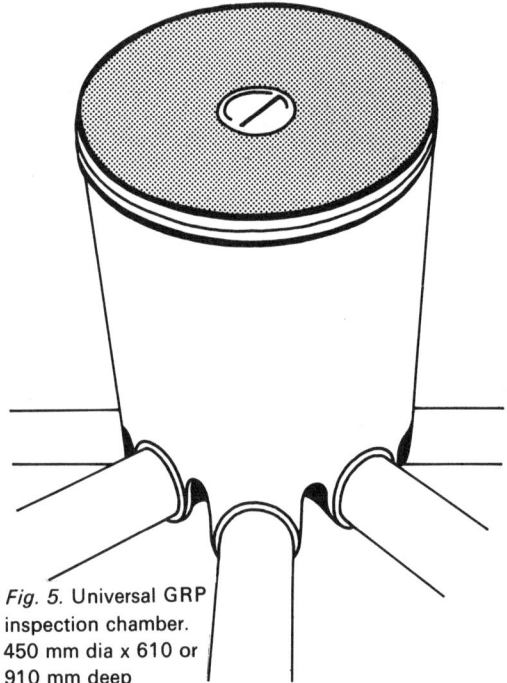

Fig. 5. Universal GRP inspection chamber. 450 mm dia x 610 or 910 mm deep

Fig. 6.

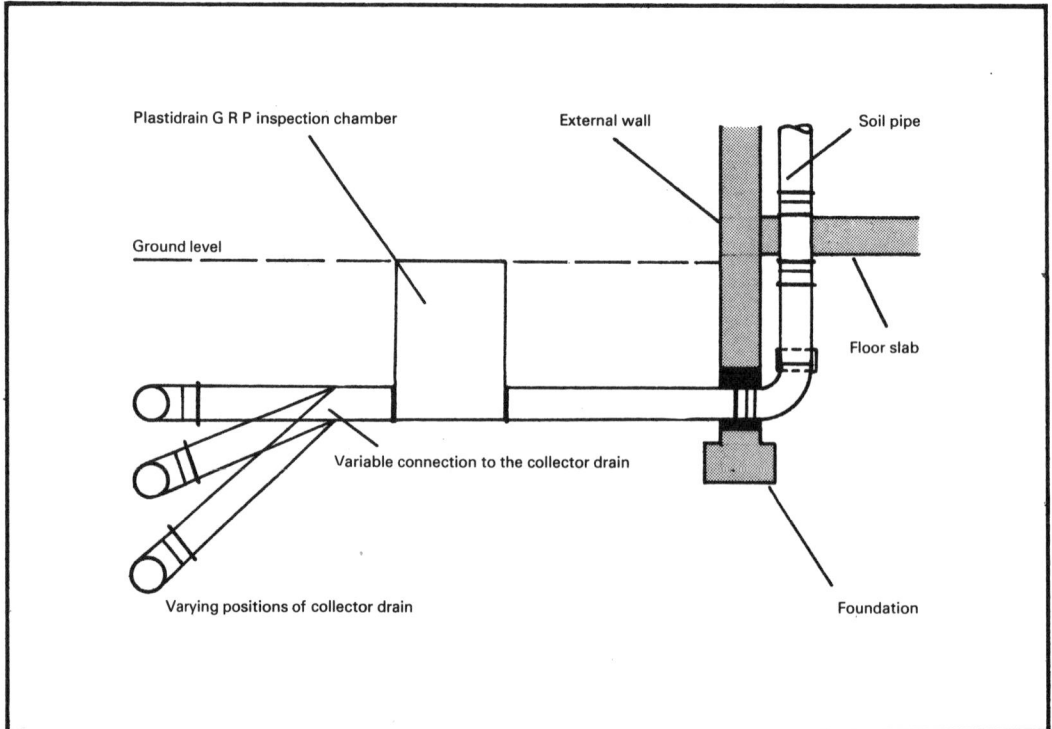

Plastidrain G R P inspection chamber

External wall

Soil pipe

Ground level

Floor slab

Variable connection to the collector drain

Varying positions of collector drain

Foundation

So that the cost of inspection chambers at the head of small branch drains may be saved, Building Regulations permit the use of Plastidrain rodding eye terminals.

Blockages

Blockages are the most common drainage faults and within the house occur most frequently in the trap below bathroom and kitchen fitments. They can often be cleared with a rubber plunger, the fitment being half filled with water, the plunger placed evenly over the drainage hole and worked vigorously up and down (see Chapters 3 and 6). If this is unsuccessful, the drain plug must be removed, care being taken to support the pipework adequately as it is unscrewed. Place a suitable container below the drain plug to catch the waste water.

Should the blockage occur in a drain it will be obvious by overflowing at the gulley, or at a rodding eye, or at the inspection chamber depending upon its situation. Modern gullies are provided with an access plug *(see Fig. 2.)* through which flexible drainage rods may be inserted. Older earthenware gullies do not have this facility and cleaning round the trap is more difficult. In either case sludge should be removed from them periodically with a scoop *(Fig. 7.)* and the grid kept clear of waste. Blockages within the inspection chamber can usually be cleared with a stick, care being taken not to damage the·pipework or in the case of bricked chambers the 'benching' around the open pipes in the bottom. Blockages at some distance along the pipe can only be freed with the aid of drainage rods and fittings, which are often available on hire.

The location of such a blockage may be determined by examining the inspection chamber(s). If there is only one and it is clear, then the blockage must be between it and the house, but if it is overflowing then the blockage may be at its outlet or beyond. Where there is more than one chamber the blockage will be above the first chamber to be found empty.

Before rodding the drain, the outflow pipe should have a piece of wire netting placed across it to catch the solids which are released, allowing the water to pass. Rodding should then begin from the chamber above using the rubber plunger head. This should be screwed to the end of the first rod and inserted into the pipe. As the rod is pushed in, further sections must be screwed on until the obstruction is reached. *At no time should the rods be turned anti-clockwise or they may become unscrewed.* Once the head of the rod touches the blockage it should be pressed firmly against it until it is released. Should this fail an alternative head should be tried — a hook, a scraper and a screwhead are usually included. The screw is designed to screw into blockages of rag or paper so that they may be withdrawn, again — *do not turn anti-clockwise.* Once a blockage has been released the system should be thoroughly flushed out and where necessary disinfected. If all attempts fail to free the obstruction, replacing the blocked section must be considered. Its location can be traced by laying the length of rods on the ground above the run. It is then necessary to dig out a trench taking care not to disturb any other services which have been laid underground. Debris from the trench should be thrown sufficiently clear to avoid running back. It must be possible to work in the trench, which if necessary, because of depth or soft soil, may have to be shored with timber. Any infilling from around the pipe should be placed separately for replacement.

Fig. 7. Tools for cleaning out gullies

Stick to which tin is nailed

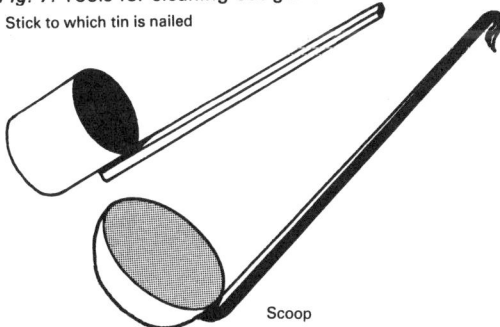

Scoop

Replacing a Pipe

The replacement of a pipe may also be necessary if it is fractured, and tracing a leak is rather more difficult. An expanding drain stopper is required *(Fig. 8.)*, so that the point furthest from the house may be plugged and the system completely filled with water, any trapped air being allowed to escape through breather tubes inserted round the bend of gullies, or by removing the access plug in modern plastic ones. A leak would be indicated if there is an appreciable fall in the water level after about an hour. Actually finding the leak may require extensive digging and inspection particularly if the drain runs under paths etc. There may be obvious points to inspect first, such as where it is clear that there has been recent subsidence, or where heavy traffic has crossed the line of the drain.

Once the fault has been located, be it a blockage or fracture, if possible the system should be plugged temporarily above this point, so that it is not flooded by the accidental discharge of waste water, and the pipe broken out taking care not to damage the ends of those pipes to which it is connected.

Earthenware pipes are easily broken up with a hammer; pitchfibre or plastic may have to be sawn. Once laid, clay pipes are rigidly joined at the socket with mortar, a tarred hemp gasket being first rammed into the joint to align the pipes and prevent mortar escaping into the bore. Alternatively they may be purchased with a plastic coating so that the joint is sealed with a rubber ring as the spigot and socket are pushed together. *(Fig. 9.)* This leaves the system flexible and less likely to fracture.

Because of the slack fitting of the joint in plain clay pipes, they are often used to repair faults as they can be eased into place *(Fig. 10.)*, the joints then being made up with gasket and mortar (one part cement to two or three of sand). It is advisable to test the system once a repair

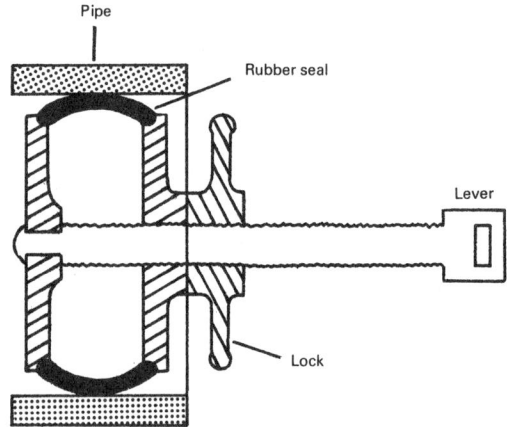

Fig. 8. Drain stopper – expanded

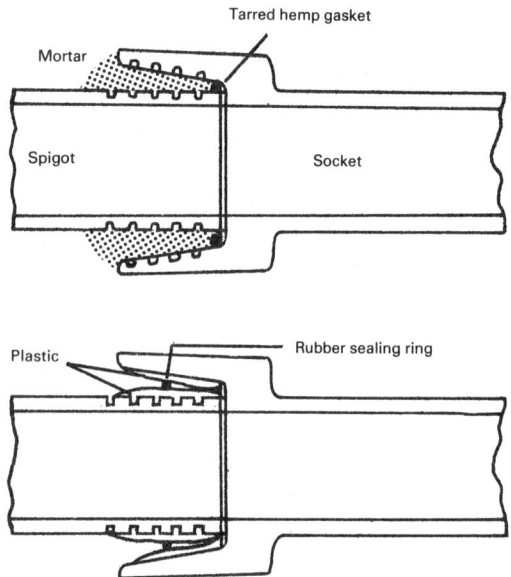

Fig. 9. Joints in clay pipes

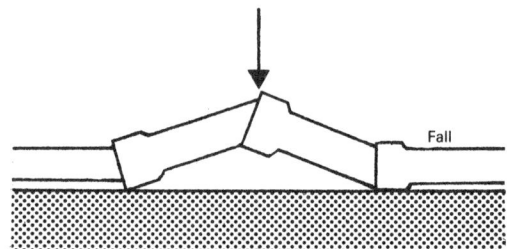

Fig. 10. Inserting new pipes

has been made before filling in the trench. Once this has been done, the infilling should be carefully packed around the pipes, and the soil then replaced in layers each being trodden down as the work progresses.

As indicated previously pitchfibre and plastic pipes may be sawn to length; if clay ones have to be cut they should be placed on soft soil and nicked round and round very carefully with a sharp cold chisel and hammer until they part. Both pitchfibre and plastic pipes may have push-fit joints sealed with a rubber ring, or plastic ones may be solvent-welded (see Chapter 11). Adaptors are also available for joining plastic pipes to clay or cast-iron systems.

16 General Care and Maintenance

Your plumbing system gives a lot of useful service generally with the minimum of trouble. This can be prolonged by taking a few simple precautions and giving the system a little thoughtful care. More than one member of the family should be familiar with its functioning, and it is sound practice to label important features, such as stopcocks and wheel valves and the supply which they control.

Blockages of traps and overflows in sinks, washbasins and baths resulting from an accumulation of grease and sediment are commonplace. The former can often be cleared with the force cup (see Chapter 3). The fitment should be partially filled with water, the cup placed over the waste pipe and the handle worked vigorously up and down. Should this fail, place a bucket below the trap and unscrew its plug, taking care to avoid damage to, or loss of the sealing washer. Then clear out each side of the trap with a flexible wire. A simple coil made by winding the wire round a length of thin dowel is ideal; this will pull out the blockage and the trap should then be flushed with hot water before the plug is replaced. Finally run more water through the trap to check that it is really clear and to remake the seal. Now ensure that the plug is not leaking and dispose of the waste water down the drain. Flexible expanding curtain wire is ideal for clearing overflows if they are found to be blocked. Simple blockages of this nature can often be avoided by regular use of cleaning fluids applied to sinks and washbasins. Similarly, cleaning fluid should be applied regularly to the WC since all pipes which discharge waste water are unhygienic. Wear rubber gloves when working on them and then wash gloves and hands thoroughly when the job is completed.

Take care not to drop bottles or jars from the bathroom cabinet into the washbasin. Those of vitreous china chip or crack too easily. Similarly avoid the use of abrasive cleaners on baths, washbasins and plated taps or other metal fittings. Hot water and household liquid cleaners are preferable.

Traces of corrosion on metal fittings can be removed with a fine abrasive cleaner, and buffing with a wax-based polish will delay further spread.

A fine slurry will often accumulate in cold-water cisterns. The opportunity to mop this out with a cloth should be taken when the water is turned off at any time, and the cistern drained. It is not advisable to scrape galvanised cisterns or holes may develop because of more rapid corrosion.

Minor leaks in threaded unions can usually be stopped by giving the nut a further half turn, ensuring adequate support is given to the rest of the pipework or fitment. Failing this, turn off the supply, drain the pipe and remake the joint with PTFE tape. Take at least two turns of tape clockwise round the male thread, pulling taut so that the tape moulds itself to the shape of the thread.

Temporary repairs to small pinhole leaks which occasionally occur in pipes can be made by binding the pipe with self-adhesive plastic tape and clamping with a small jubilee clip until the defective tube can be cut out and replaced.

Make sure that the outdoor stop tap is always accessible. Clean deposits of dirt from round the lid which covers it, and in periods of frost apply salt to prevent it from freezing fast.

It is becoming increasingly advisable to be economical in the use of water. It is a commodity for which there is a growing demand – on average 150 litres per person per day.

Avoid dripping taps and waste from overflows. It has been calculated that one drip per second wastes 1,416 litres per year.

Finally, remember that all water authorities have bye-laws relating to contamination, misuse and waste of water. These bye-laws also define the kinds of pipes and fittings which may be used in their particular area. They are regulations with which your plumbing system, and any modifications to it, must conform.